国家科学技术学术著作出版基金资助出版

熔体静电纺丝

——生产超细纤维的绿色方法

Melt Electrospinning

A Green Method to Produce Superfine Fibers

刘勇（Yong Liu）
李凯丽（Kaili Li）
[印　度] 阿泽（Mohamedazeem M. Mohideen）
[新加坡] 席瑞思（Seeram Ramakrishna）
著

·北京·

ACADEMIC PRESS
An imprint of Elsevier

国家科学技术学术著作出版基金资助出版

Melt Electrospinning
A Green Method to Produce Superfine Fibers

熔体静电纺丝——生产超细纤维的绿色方法

刘　勇（Yong Liu）
李凯丽（Kaili Li）　著
[印　度]　阿泽（Mohamedazeem M. Mohideen）
[新加坡]　席瑞思（Seeram Ramakrishna）

内容简介

熔体静电纺丝是纳米纤维制造的新兴技术之一，其最大的特点是不用添加溶剂，具有无毒、环保、安全、经济等方面的优势。在生物医学、药物控释、组织工程等方向具有广阔的应用前景。

本书基于作者多年的原创科研成果，对熔体静电纺丝技术做了全面的总结，共四部分。第一部分介绍了熔体静电纺丝的发明，包括离心熔体静电纺丝和向上熔体静电纺丝的独立发展，分别对两种方法的产率和纤维直径进行了优化。第二部分介绍了熔体的静电纺丝以及利用不同聚合物和自行设计的装置测试纤维性能的方法。第三部分介绍耗散粒子动力学模拟，这种模拟技术是模拟纺丝过程中分子链结构和取向的一种方法。第四部分介绍了离心熔体静电纺丝的原理、方法及改进措施。

本书不仅适合静电纺丝研究的广大科研人员阅读，同时还可供燃料电池、锂电池、太阳能电池、水过滤、空气过滤、血液过滤、组织工程、载药缓释、癌症检测、介入治疗支架、人造血管、金属吸附等可能用到纳米纤维的广大领域的科技工作者、研究生、企业管理人员参考。

图书在版编目（CIP）数据

熔体静电纺丝：生产超细纤维的绿色方法＝Melt Electrospinning：A Green Method to Produce Superfine Fibers：英文/刘勇等著. —北京：化学工业出版社，2021.11
ISBN 978-7-122-39799-7

Ⅰ.①熔…　Ⅱ.①刘…　Ⅲ.①熔融纺丝-英文　Ⅳ.①TQ340.642

中国版本图书馆 CIP 数据核字（2021）第 172058 号

责任编辑：吴　刚　李玉峰　　　　封面设计：关　飞
责任校对：王　静

出版发行：化学工业出版社（北京市东城区青年湖南街 13 号　邮政编码 100011）
印　　装：北京虎彩文化传播有限公司
710mm×1000mm　1/16　印张 $13\frac{3}{4}$　字数 177 千字
2022 年 1 月北京第 1 版第 1 次印刷

购书咨询：010-64518888　　　　售后服务：010-64518899
网　　址：http://www.cip.com.cn
凡购买本书，如有缺损质量问题，本社销售中心负责调换。

定　　价：**298.00 元**　　

Contents

About the authors

Yong Liu is an associate professor of the College of Materials Science and Engineering at Beijing University of Chemical Technology (BUCT). He is a director of the polymeric Nano Composite group at BUCT. He received his Ph.D. from the Institute of Chemistry, Chinese Academy of Science, in 2005. Prof. Liu worked as a visiting associate professor at Cornell University in 2011–2012. He had committed his research work to an project, as a post-doctoral researcher in Tsinghua University for more than 2 years. He is currently supervising masters, international Ph.D. students, and postdoctoral fellows. His research interest is focused on the application of highperformance engineering plastics, performance advancement of rubber products, spinning of special functional fibers, preparation of nanofibers by electrospinning, decomposing formaldehyde in the dark and at room temperature, preparation of membranes for fuel cells especially in polymer electrolyte membrane (PEM) fuel cells, and the application and preparation of nanoparticles. He has published 103 articles in peer-reviewed journals and presented about 47 oral and poster presentations in national and international journals. Prof. Liu has applied for 83 patents, of which 47 are currently issued. He serves as a reviewer for more than 40 journals such as Polymer, RSC Advances, Polymeric Engineering and Science, Journal of Applied Polymer and Science. Dr. Prof. Liu is a member of the Royal Society of Chemistry, a senior member of the Chinese Society for Composite Materials, a member of the American Chemical Society, and a member of the Chinese Chemical Society. He was awarded by the State Council, with the National Award for Science and Technology Progress (Grade 2), for his contribution to polymer materials science. As a famous scientist in nanofibers, he has been interviewed by and appeared on CCTV, BTV, Dragon TV, and many other television programs and in newspaper articles.

Kaili Li is currently a postgraduate at Beijing University of Chemical Technology. Her research focuses on the movement of molecular chains, the fibers produced by centrifugal melt electrospinning, and dissipative particle dynamics simulation. She has published several articles in peer-reviewed journals.

Mohamedazeem M. Mohideen is pursuing a Ph.D. at Beijing University of Chemical Technology (BUCT). He is doing his research work under the guidance of Prof. Yong Liu. He completed his master's degree in Materials Science from Anna University, India, in 2018. He finished his master's dissertation on quantum dots with the cooperation of the Department of Metallurgical and Materials Engineering - IIT Madras. Currently, his research work is focused on the improvement of catalyst activity in proton exchange membrane fuel cells.

Seeram Ramakrishna, FREng, FBSE, is the Director of the Center for Nanofibers and Nanotechnology at the National University of Singapore (NUS). He is regarded as the modern father of electrospinning and applications. He is a highly cited researcher in material science and in 2004 he was recognized as being the world's most influential scientific mind by Thomson Reuters. He has coauthored over 1000 SCI-listed journal papers, which have received ~86,000 citations and ~140 H-index. He is an elected fellow of the UK Royal Academy of

Engineering (FREng), Biomaterials Science and Engineering (FBSE), the American Association of the Advancement of Science (AAAS), and the American Institute for Medical and Biological Engineering (AIMBE). He has received such prestigious awards as the IFEES President Award—Global Visionary, the Chandra P. Sharma Biomaterials Award, the Nehru Fellowship, the LKY Fellowship, the NUS Outstanding Researcher Award, the IES Outstanding Engineer Award, and the ASEAN Outstanding Engineer Award. His academic leadership includes being Vice-President for Research Strategy, Dean of the Faculty of Engineering and Director of Enterprise at the National University of Singapore, and Founding Chairman of the Solar Energy Institute of Singapore. His global leadership includes being the Founding Chair of the Global Engineering Dean's Council and Vice-President of the International Federation of Engineering Education Societies. He received a Ph.D. from the University of Cambridge, UK, and General Management Training from Harvard University, USA. He is an advisor to universities, corporations, and governments around the world. Prof. Seeram serves as an editorial board member for 12 international journals.

Preface

At the point when we started to write this book, it was an exciting and great challenge for us to condense our past several years' research work and experience into a few chapters. The evaluation of nanoscience and technology results in the innovation of potential applications in various fields. Under the influence of nanotechnology, nanofibers are an attractive application in our scientific world. Energy production is more efficient nowadays; and nanofibers can be utilized for the production of batteries and fuel cells in energy storage fields. Similarly, it can have uses in various fields such as drug delivery, tissue engineering, and so on. However, the production of such fine nanofibers from micro and nano scales has increased interest in the electrospinning process. Therefore, for several past decades, solution electrospinning has mostly been used in the fabrication of nanofibers and the number of related research and conference papers has increased. However, the major drawback is that solution electrospinning is not environmentally friendly due to residual toxic solvents present in the fibers.

The German chemist Manfred Eigen said "*A theory has only the alternative of being right or wrong. A model has a third possibility: it may be right, but irrelevant.*" Just like what he said, solution electrospinning may produce fibers with various advantages but it has certain drawbacks, which are harmful to humankind especially in certain applications. To overcome this, the novel and green approach of melt electrospinning has been developed. In this book, the main focus is on melt electrospinning, but there is only a limited amount of research conducted on this subject. The development of melt electrospinning devices is a hot topic all over the world. This book summarizes the different designs of melt electrospinning devices and their working processes, with the help of systematic and clearly presented diagrams. The main aim of this book is to break the controversies and limitations surrounded by melt electrospinning and to motivate up-and-coming researchers to open the gates for its future prospects.

Before going through this book, we hope that the reader has some knowledge in polymer physics and simulation studies. This book will share some important prospects and ideas with readers about the evaluation of melt electrospinning equipment designs, factors affecting fiber diameter, simulation studies on fibers, the design of centrifugal melt electrospinning, and the application of nanofibers in drug-delivery systems.

Since it is co-published with Elsevier Inc., the present edition follows the typesetting of Elsevier's edition, including, but not limited to, the fonts, sizes, subscripts, superscripts, normal or italic letters, reference, as a courtesy.

Yong Liu
Kaili Li
Mohamedazeem M. Mohideen
Seeram Ramakrishna

Acknowledgments

After the completion of this book, it is now time to acknowledge the individuals who helped us during the period of writing this book. The chapters in this books are based on the first author's past research collaboration experience with various researchers, and various individuals patents are used as a reference to highlight the growth and possibilities of greener production of nanofibers via melt electrospinning. However, it is quite difficult to express our regard to each individual person, but we would like to extend our gratitude to those who laid a pillar at various stages which has enabled us to finish this work as early as possible.

The authors would like to thank the colleagues for taking part in the research work and their valuable suggestions which lifted the standard of every chapter. We would like to extend our thanks to Beijing University of Chemical Technology (BUCT) for providing us with a wonderful working atmosphere with highly advanced laboratories, in which to do innovative research work. And we would also like to thank the National Nature Science Foundation of China for their continuous funding support of our research.

The suggestions from the reviewers of this book are very much appreciated and their valuable comments helped a great deal. The development and preparation of the book were facilitated by a number of dedicated people at Elsevier and Chemical Industry Press. We would like to thank all of them, with special mention going to Gang Wu of Chemical Industry Press and Dr. Glyn Jones of Elsevier. It has been a great pleasure and fruitful experience to work with them in the preparation and publishing of this book.

We would also like to thank group members of the Polymeric Nano Composite Laboratory and each of the graduate students from the research laboratory. Their contribution to this book was highly appreciated and most valuable. The first author would like to express his thanks to the students who worked in all the chapters to make this book possible. Mr. M. Mohamedazeem has written an introduction and added additional contents in each chapter. Miss Kaili Li compiled the entire contents of the nine chapters. Prof. Yong Liu and Prof. Seeram Ramakrishna designed the book and guided the writing process.

Yong Liu
Kaili Li
Mohamedazeem M. Mohideen
Seeram Ramakrishna

Chapter 1

Development of melt electrospinning: the past, present, and future

Chapter outline

In this introductory chapter, some historical events of melt electrospinning are discussed. In-depth analysis of melt electrospinning is summarized in later chapters. This chapter is mainly focused on the basics, advantages, and limitations of both the solution and electrospinning processes. Finally, a brief summary of the following chapters is given.

In 1936, Charles Norton, a physicist from the Massachusetts Institute of Technology, was the first to describe melt electrospinning for the production of fiber by an air blast in his patent [1]. In 1938, Games Slayter described in his patent the formation of glass fibers by an air blast [2]. Almost 50 years later, the first scientific paper was published, by Larrondo and a co-worker John Manley, in 1981 using polyethylene and Nylon 12 [3,4]. They broke down the controversies that not only fluids and polymer solutions, but also polymer melt can form a jet by the influence of an electrostatic field. After two decades, in the early 21st century, a second scientific paper on melt electrospinning was published by Reneker and Rangkupan [5]. Over the last few decades, many polymer materials have been melt electrospun by melt electrospinning techniques, with temperature ranges up to 380°C. But since then only a limited number of works have been reported. Before going too in-depth into the melt electrospinning device and its limitations, we first discuss some of the basics of electrospinning.

1.1 Electrospinning

Firstly, what is electrospinning? Electrospinning is a process of producing nanofibers with a diameter ranging from nanometers to a few micrometers

Melt Electrospinning. https://doi.org/10.1016/B978-0-12-816220-0.00001-4

on the application of an electric charge on a polymer melt or solution. In other words, the process of electrically forcing liquid to draw fibers is also known as electrospinning. The electrostatic attraction of liquid was first recorded by William Gilbert in AD 1600 [6]. In 1934, Anton Formhal's was the first patentee to describe the electrospinning apparatus to produce a polymer filament under the action of an electrostatic force. By knowing the basic principle of electrostatics and capillarity, one can understand the depth of the physical principle of the electrospinning process. Various types of materials can be utilized to produce nanofibers via electrospinning, such as polymers, biomaterials, ceramics, and inorganic compounds.

1.2 The working principle of electrospinning

The common electrospinning device setup is composed of a pump, a syringe, a nozzle, a high-voltage power supply, and a collector plate. A polymer is completely dissolved in a suitable solvent and fed into the syringe. A syringe pump is used to push the solution into the tip of the needle, which is placed at a certain distance from the collecting plate. Therefore, a solution droplet is formed at the tip of the needle. On application of a high-voltage power supply, an electric field is created between the tip of the nozzle and the collector. When the surface tension of the liquid droplet is overcome by the force of the electric field, the droplet is distorted, forming the so-called Taylor cone. On further increasing the voltage, the cone becomes unstable and a liquid jet emerges from the apex of the cone and evenly falls on the collector plate.

1.3 Types of electrospinning

Depending upon the state of the polymer used, electrospinning is classified into two types: (1) solution electrospinning and (2) melt electrospinning.

1.4 Solution electrospinning

Solution electrospinning is the most widely utilized technique to produce fibers with a submicron range. In solution electrospinning, the nature of the polymer is a solution, which is completely dissolved in a solvent. However, with solution electrospinning and the use of solvents, there is a series of problems, as follows.

The solvent used for spinning is very expensive and most of the toxic waste must be recovered, but within an electrospinning factory, solvent recovery is quite difficult.

Fibers in the field of biomedical applications, such as cell scaffolds, surgical procedures, etc., must ensure that these fibers are safe, but in the process of solution electrospinning residual solvent on the fiber will exist, thus limiting its applications in the field.

(1) Some polymers such as polypropylene do not have an appropriate solvent for the spinning solution at room temperature.
(2) The evaporation of the solvent during the spinning process will leave traces on the fiber, which results in the fiber surface not being smooth, or even in defects occurring, which seriously affects the strength of the fiber.
(3) The concentration of the spinning solution is very small, and the evaporation of the solvent greatly reduces the yield and wastes energy.
(4) In the solution electrospinning process, the capillary is often blocked, which affects the continuous production of fibers.

In view of the above problems, researchers have turned their attention to producing green and environmentally friendly fibers by melt electrospinning, which is discussed in detail next.

1.5 Melt electrospinning

In melt electrospinning, the polymer is melted, and used for electrospinning, to produce ultrafine nanofibers. Compared to solution electrospinning, melt electrospinning has many advantages in overcoming the problems encountered with solution electrospinning, as described above. In melt electrospinning, the name itself indicates that it does not require any dissolution of polymers in solvents. Moreover, melt electrospinning is free from toxicity, so it can be used in biomedical applications. Hence, due to the solvent-free nature of melt electrospinning, it can produce fibers from polymers such as polyethylene (PE), polypropylene (PP), and polyethylene terephthalate (PET). Melt electrospinning fabricates fibers with a diameter ranging from nanoscale to microscale. And the fiber diameter is greatly controlled by changing the parameters during the process. The electric field greatly influences the fiber diameter of electrospun polymers. When the application of the electric field increases, there is a decrease in the fiber diameter. Moreover, some studies reveal that, depending on the collector plate distance, fiber diameter can be varied. According to Yong Liu et al., [7] it is revealed that on increasing the distance between the collector and the emerging jet nozzle to some extent, there will be a decrease in the fiber diameter. This is mainly due to the fiber stretching time and the electric force influenced by the collector and the fiber. The flow rate of the fiber is another parameter to control the fiber diameter. Moreover, the viscosity of the melt is the most important parameter in the fiber diameter. As spinning temperature increases, there will be a decrease in the viscosity of the melt, which helps to produce a fiber with a smaller diameter.

In spite of all of those advantages over solution electrospinning, only a limited amount of research work has been carried out on melt electrospinning. These limitations are mainly because of some complex equipment designs. Usually, solution electrospinning equipment consists of a high-voltage power supply, a syringe with a small nozzle and the collector plate. In addition to this, melt electrospinning equipment is composed of complicated components such as (1) high spinning voltage and (2) a heating system, which is used to melt the polymers at an appropriate viscosity, with one end is connected to a spinneret to produce fiber. Moreover, there is a variety of heating sources designed by researchers, which are explained in Chapter 2.

1.6 The scope of this book

This book aims to introduce the research results of the preparation of nanofibers by melt electrospinning. It introduces the experimental technology of the melt electrospinning process and discusses the theoretical problems. It aims to stimulate the reader's interest in the preparation of nanofibers by melt electrospinning. The book is divided into four parts: Chapter 2 mainly describes the existing melt electrospinning device and the laboratory homemade melt electrospinning device, and analyzes the advantages and disadvantages of their respective devices. Chapter 3 mainly introduces the specific experiment of preparing the fiber by melt electrospinning. The experimental results are analyzed, discussed, and predicted. Chapter 4 describes melt electrospinning in a parallel electric field. Chapter 5 introduces the program for melt electrospinning by Dissipative Particle Dynamics (DPD) developed by Liu Yong, the experimental group, which is simpler than the previous DPD simulation, and the calculation of microscopic particles realized by program language C++. Chapter 6 introduces the centrifugal electrospinning equipment developed by our laboratory, which combines the advantages of centrifugal spinning and electrospinning, and eliminates the shortcomings of both, improving the spinning efficiency and fiber properties. In this chapter, the performance of the centrifugal melt electrospinning equipment is explored by studying the fiber yield, fiber diameter, and crystallinity. Finally, the equipment is upgraded according to the spinning condition. Chapter 7 mainly introduces the analysis process and simulation results of DPD simulation combined with centrifugal melt electrospinning. Chapter 8 describes the three-dimensional (3D) printing based on controlled melt electrospinning in polymeric biomedical materials. Chapter 9 shows the application of melt electrospinning in biomedical applications.

References

[1] Norton CL. Method of and apparatus for producing fibrous or filamentary material. 1936. US Patent 2048651.

[2] Slayter G. Method and apparatus for making glass wool. USA. 1938.

[3] Larrondo L, John Manley RS. Electrostatic fiber spinning from polymer melts. III. Electrostatic deformation of a pendant drop of polymer melt. Journal of Polymer Science Part B: Polymer Physics 1981;19(6):933–40.
[4] Larrondo L, John Manley RS. Electrostatic fiber spinning from polymer melts. II. Examination of the flow field in an electrically driven jet. Journal of Polymer Science Part B: Polymer Physics 1981;19(6):921–32.
[5] Doshi J, Reneker DH. Electrospinning process and applications of electrospun fibers. In: Industry Applications Society Meeting. IEEE; 2002.
[6] Zeleny J. Instability of electrified liquid surfaces. Physical Review 1917;10(1):1–6.
[7] Liu Y, Deng R, Hao M, et al. Orthogonal design study on factors effecting on fibers diameter of melt electrospinning. Polymer Engineering and Science 2010;50:2074–8.

Chapter 2

The device of melt electrospinning

Chapter outline

2.1 Introduction

With the development of nanoscience and technology, the production of nanofibers has increased interest toward the electrospinning process in both academic and industrial applications. Because of the lack of solvent in the process of melt electrospinning, it is not necessary to consider the recovery of the solvent and the toxicity of the solvent. It has the characteristics of economy, safety, and environmental protection. However, due to these advantages over solution electrospinning, melt electrospinning has attracted attention from researchers to produce green nanofibers. It has been widely used in high-performance nonwoven, biomedical, and high-efficiency filter materials. Melt electrospinning is a method for preparing ultrafine fibers by forming a charged polymer melt in an electrostatic field to form a jet.

2.2 Conventional melt electrospinning devices

The melt electrospinning device is more complex than the solution electrospinning device. Melt electrospinning devices currently built by various research institutions have been widely modified to make the design uncomplicated and for commercial suitability [1]. As shown in Fig. 2.1 an electric field is supplied between the tip of the spinneret melt and the receiving device,

Melt Electrospinning. https://doi.org/10.1016/B978-0-12-816220-0.00002-6

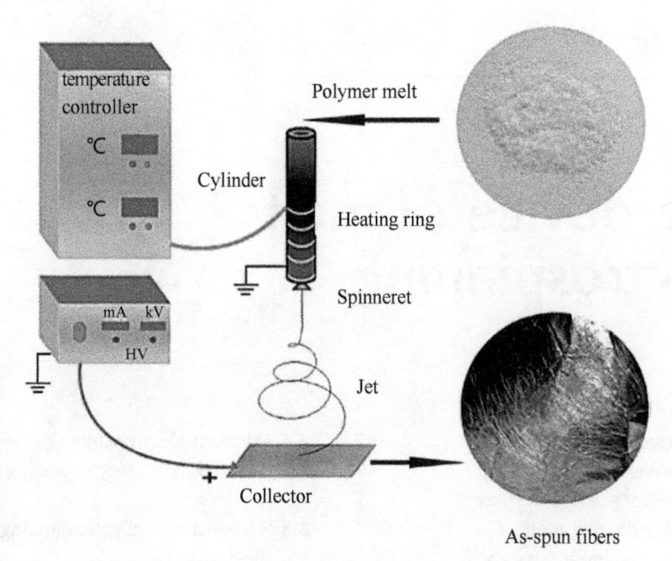

FIGURE 2.1 Schematic of the melt electrospinning device [2].

for the electrostatic force to overcome the surface tension of the melt, and enable superfine fiber to be drawn. Melt electrospinning devices are currently built by various research institutions and there is no recognized mature equipment. In our research laboratory, we designed a melt electrospinning device for the fabrication of poly(lactic acid) fibers. We invented a cone-shaped spray head instead of the typically used capillary tube model. The following sections discuss several types of typical devices.

2.3 Laser heating melt electrospinning devices

To overcome the bottleneck of electric discharge in conventional melt electrospinning, a CO_2 laser beam heating device is first introduced by Nobuo Ogata et al. [3–5] (shown in Fig. 2.2). The CO_2 laser beams, with a 5 mm diameter, irradiate from three directions on a polymer rod, such as poly(lactide), and evenly melt. Continuous activation of the laser beam on the sample gets melts by applying a high voltage between the melted parts and the rotating disc, resulting in the formation of a Taylor cone, from which a fine diameter of polylactic acid (PLA) fiber, of less than 1 μm, is drawn. Experiments show that the polymer with polar groups, such as PLA, polyethylene-co-vinyl alcohol (EVOH), polyamide, using the device melt electrospinning, can be nanofiber. And polymers without polar groups, such as polypropylene and polyethylene, are not nanograde fibers.

However, the major problem is that the mass production of fibers is not possible using this device. To overcome the above problem, in 2010, Shimeda et al. designed and used a line-like laser heating device, instead

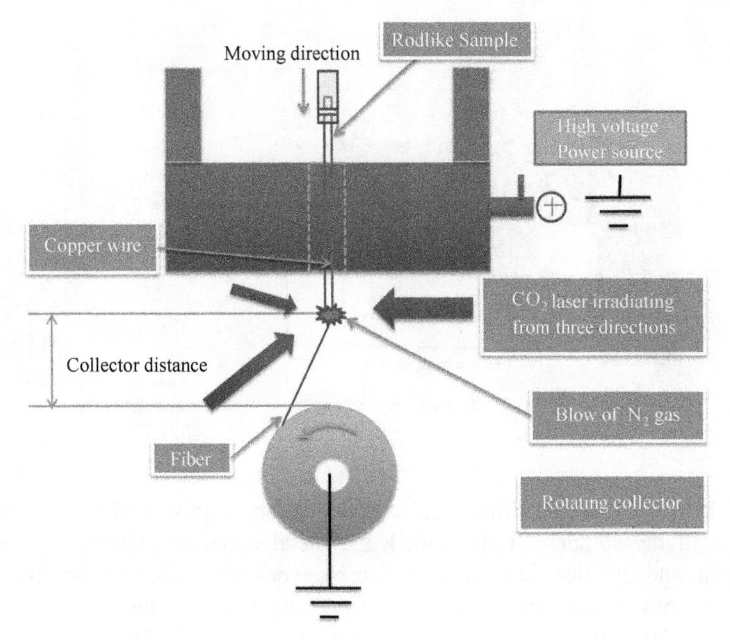

FIGURE 2.2 Schematic diagram of a laser heating melt electrospinning device [3].

of a rod sample polymer sheet. Polyethylene-co-vinyl alcohol (EVOH) and Nylon 6/12 can be easily melted by using laser melting devices. For further improvement in fiber diameter, in 2012, the same group combined PLA and EVOH and produce a bundle fiber of a diameter about 400 nm using a spot laser melting device. This is low when compared to fiber produced from solution electrospinning (about 800 nm). The use of laser heating, having the advantages of fast heating, low energy consumption, and noninterference with a high-voltage electrostatic loading system can be used as a better experimental device. However, the temperature and viscosity of the instantly heated polymer material cannot be effectively controlled. The high cost and safety of the laser are also problems that the researcher needs to focus on, and so the technology may be limited on the road to industrialization.

2.4 Screw extrusion melting electrostatic spinning devices

Commonly used in the melt electrospinning devices are a glass syringe, metal electrode, polymer, and heating wire, which are all isolated from each other, without electrical interference. However, in the screw extruder melt electrospinning device (shown in Fig. 2.3), the heating part of the screw extruder is metal and is directly connected to the motor and the power supply. To avoid a

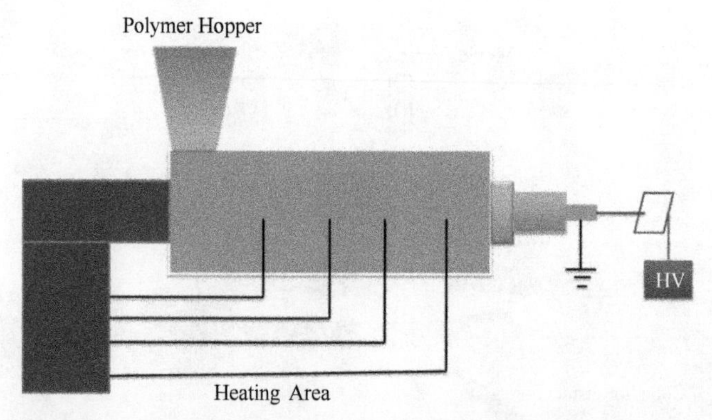

FIGURE 2.3 Schematic diagram of screw extrusion melt electrospinning [1].

short circuit in the machine, the extruder must be grounded to ensure no charge. If the spinneret is grounded, the metal collector plate is negatively charged, and the electric field force can be generated. When the voltage reaches the critical value, the electric field force can overcome the surface tension and viscoelasticity of the melt, so that the fiber can fly out of the spinneret [6]. During the spinning process, as the material continues to deliver, the diameter of the jet continues to decrease until the viscosity once again overcomes the electric field force to cure the jet stream. Compared with other melt electrospinning methods, the use of the device in the electrospinning process does not appear to yield instability, which may be due to the high viscosity of the melt and a shorter spinning distance.

2.5 Electromagnetic spinning devices for vibration

The vibrating electrospinning device forms a jet by causing the liquid in the solution tank to fluctuate and overcome its own gravity. Under the action of the electric field force, nanofibers are generated on the receiving plate [7]. The major modes of vibration are ultrasonic vibration, mechanical vibration, etc. Wan Yuqin [8] and group, of Donghua University, studied the ultrasonic vibration melt electrospinning device (shown in Fig. 2.4). The high-frequency shear vibration force field generated by ultrasonic waves is parallel to the steady-state shear flow of the fluid, reducing the entanglement between the molecular chains, reducing the resistance of the fluid flow, improving the fluidity of the fluid, and reducing the apparent viscosity of the fluid. At the same time, ultrasonic action on the polymer fluid will cause local degradation, the molecular chain is short, the degree of entanglement of the chain is reduced, and the molecular chain of the transition or slip becomes easy. Therefore, the mobility becomes better, achieving the effect of polymer viscosity reduction. Ultrasonic, temperature field, and electrostatic field of

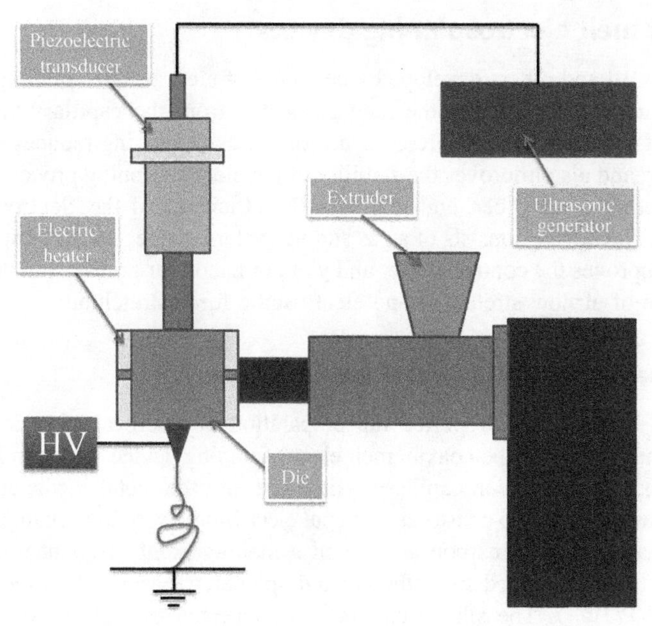

FIGURE 2.4 Schematic diagram of a vibrating temperature electrostatic discharge device [8].

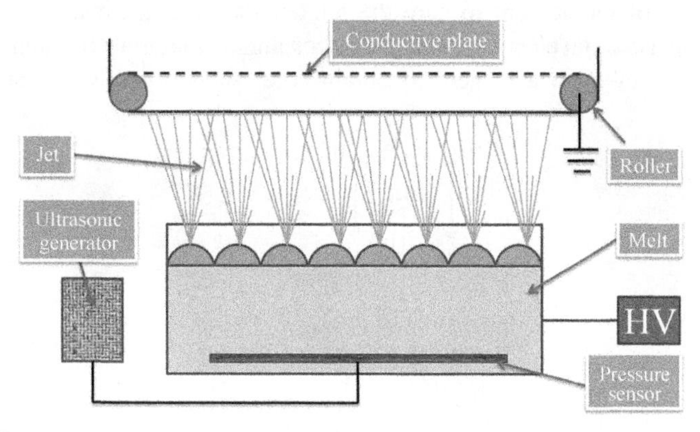

FIGURE 2.5 Schematic diagram of an upwardly vibrating melt electrospinning device [9].

multifield coupling is a collection of the ultrasonic and electrostatic fields, temperature field, and electrostatic field, coupling two advantages. In short, due to the role of ultrasound, the melt of the spinnability has been greatly improved. He Jihuan et al. [9] invented an upwardly vibrating melt electrospinning device (as shown in Fig. 2.5). The production efficiency is high, and it enables large-scale production of nanofibers to meet the industrial applications and daily necessities of nanofiber mat demand.

2.6 Air melt electrospinning devices

Chi Lei [10] and others developed a new type of electrostatic spinning device (shown in Fig. 2.6). When the melt is ejected from the capillary under the action of the electric field force, the use of hot air stretching reduces the fiber diameter, and also improves the stability of the electrospinning process, guides the direction of the fiber, and improves the efficiency of the electrospinning process. The device consists of a gas supply system connected to the spinneret, which improves the controllability and yield of nanofiber preparation under the influence of airflow stretching and electrostatic force stretching.

2.7 Coaxial melt electrospinning devices

Jesse T. McCann [11] reported the preparation of microcapsules and phase change nanofibers by the coaxial melt electrospinning device shown in Fig. 2.7. A polymer-coated silicon capillary is inserted into the metal needle at the end of the plastic syringe to construct a coaxial electrostatic sprinkler. In this way, an alkane having 16–20 carbon atoms and a melting point of about room temperature, can be ejected from the coaxial spinneret together with the solvent (shell PVP/TiO_2). The silicon capillary is connected to a glass syringe with insulated heating wire. The temperature is regulated by the temperature controller. In the spinning process, the cooling of the spray mainly depends on the evaporation of the solvent to cure the molten state present inside. Similar to conventional coaxial electrospinning, electrospinning materials (molten or solid) must be insoluble in a solvent to obtain core–shell fibers. Core–shell-type

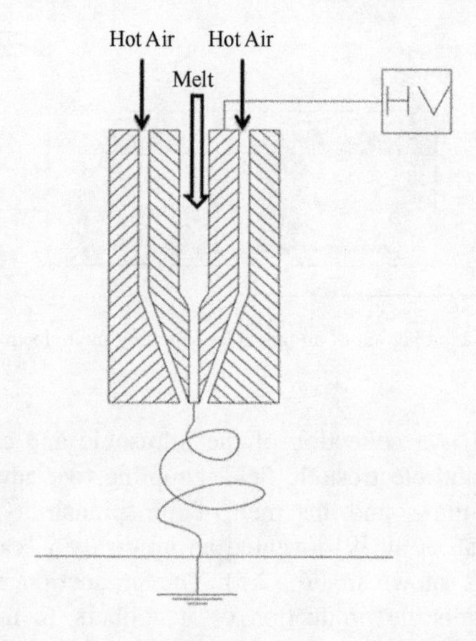

FIGURE 2.6 Schematic diagram of airflow melt electrospinning [10].

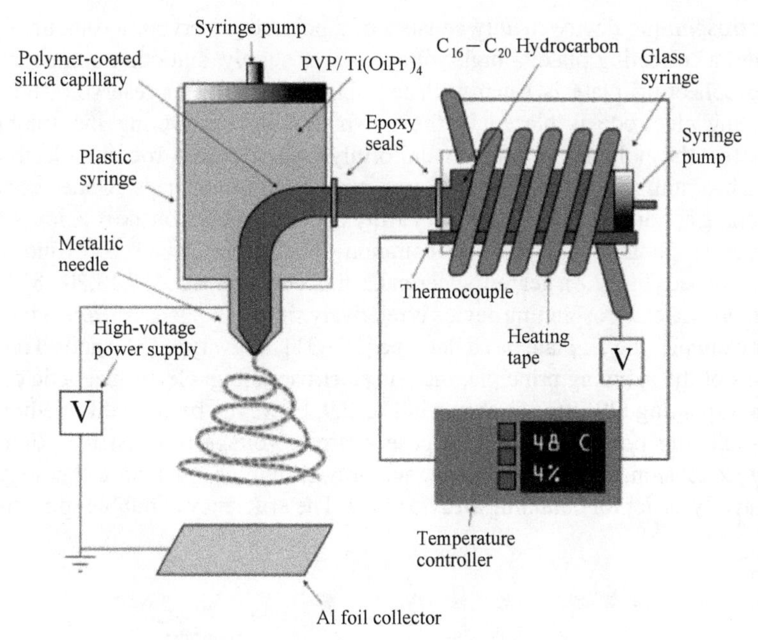

FIGURE 2.7 Schematic diagram of coaxial electrostatic spinning apparatus [11].

nanofibers are widely used, such as in the preservation of unstable biological agents, to prevent the decomposition of unstable compounds, the sustained release of molecular drugs, the construction of tissue engineering scaffolds, conductive nanowires, chemical protective clothing, etc. [12].

2.8 Upward melt electrospinning devices

The method of depositing the electrospinning fiber product in the lower part of the collecting plate may be referred to as upward electrospinning. A free surface electrospinning technology evolved by conventional electrospinning technology, it has been advantageous in preventing the spinning needle from becoming clogged, while greatly improving the fiber yield. Since the development of nano spider prototypes [13], needle-free electrospinning has been increasingly favored by researchers. A great deal of research and exploration has been done on the electrostatics of upward electrospinning, including rotary bath electrode electrospinning, free surface electrospinning, and bubble spinning [14–24]. Through the unremitting efforts of many scientific research workers, upward electrospinning technology has been gradually commercialized and mass production of nanofiber membrane was achieved [25].

In solution electrospinning technology, solution bath rotary electrode electrospinning, free surface electrospinning, and bubble electrospinning technology are well known. Among them, the solution bath rotary electrode-type electrospinning device mainly consists of a polymer reservoir, a rotating electrode, a collecting plate, a high-voltage power supply, and other components. The collecting plate is usually placed above the polymer reservoir and the rotating electrode is placed in the polymer reservoir. During the spinning process, the polymer solution is uniformly coated onto a rotating electrode, which is uniformly rotated and forms numerous helium taps. Under certain spinning conditions, the formation of a Taylor cone depends on the rotation of the electrode geometry. The most common rotating electrode is cylindrical, disc-shaped (Fig. 2.8), geared, spherical, spiral coil type, etc. [17,18,20,28]. The free surface electrospinning device is relatively simple, but it has different forms. For example, the jet plate has a flat type [29–31], a saw type [31], and so on. In terms of the spinning principle, there is electrospinning, electromagnetic combined spinning [29], etc., as shown in Fig. 2.9. However, because this method in the spinning process needs to impose a strong voltage to resist the polymer surface tension, viscous resistance, and gravity, bubble spinning technology is relatively easier for obtaining a Taylor cone. The efficiency of bubble spinning is

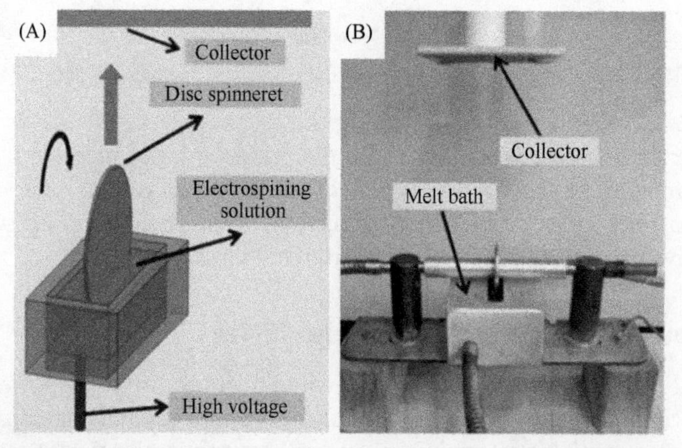

FIGURE 2.8 (A) A schematic diagram of disk electrode type solution electrospinning [26], (B) The disk electrode type melt-electrospinning device [27].

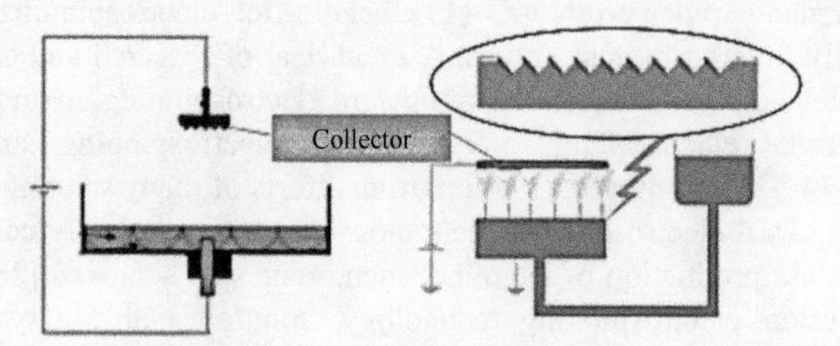

FIGURE 2.9 A schematic diagram of electromagnetic [29] and sawtooth type [31] solution electrospinning.

high and the yield is large. However, due to factors such as the yield, uniformity, and size of the spinning fiber, the influence of the number, distribution, size, and solution viscosity of the bubble is relatively large. However, to effectively control these parameters, further exploration and research are needed.

The research into solution electrospinning is more abundant. From a past report [32], Fig. 2.8B shows rotating disk electrode melt electrospinning, which is similar to the principle of the rotary bath electrode electrospinning, as shown in Fig. 2.8A, where the polymer solution is introduced into the melt bath and rotating electrode. On application of a high-voltage power supply, polymer solution is uniformly heated by a rotating electrode in the melt bath and produces the electrospun fiber. Although the method can efficiently produce continuous ultrafine fibers (2 μm), due to the applied voltage at about 60 kV, it is much higher than the conventional melt electrospinning voltage. On the one hand, it increases the risk of the experimental process, and on the other hand the rotating electrode needs to be semisubmerged in the polymer melt pool where the spinning will lead to an unnecessary waste of material.

The self-designed melt electrospinning machine is shown in Fig. 2.10. The heating ring, plunger, etc., are fixed on the barrel according to the position in the figure, and the collecting plate is fixed on the distance adjusting bracket. The barrel and the pitch bracket are fixed on the base, and the pusher is pressed against the lower end of the plunger. The temperature control box and the high-voltage power supply are, respectively, connected to the heating coil and

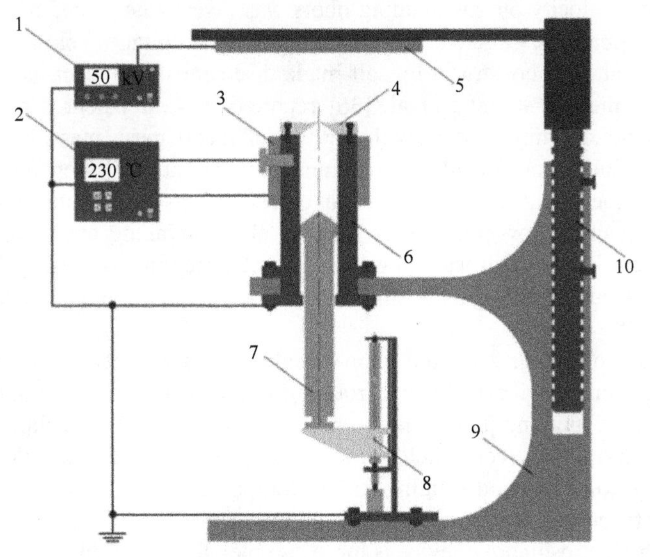

FIGURE 2.10 A schematic diagram of an upward melt-electrospinning machine. 1. High-voltage power supply; 2. temperature controller; 3. heating ring; 4. spinneret; 5. collector; 6. feed cylinder; 7. plunger; 8. propeller; 9. base; 10. pitch control device[33].

the collecting plate. In the experiment, the temperature is set by the temperature adjustment button on the temperature control box. After the set temperature is reached in the barrel and stabilized for 10 min, the material is added from the upper end of the barrel and the spinneret is screwed to the upper end of the barrel. When the material in the barrel is fully melted, the pusher is opened, and pushed the plunger upward at a constant, minute speed, and the material is extruded out of the spinneret. At this time, the high-voltage power supply is turned on, and parameters such as the spinning voltage are set. It is then observed that a high number of Taylor cones were formed above the spinneret, the cone tips moved toward the collecting plate, formed fibers, and were deposited on the lower surface of the collecting plate.

2.9 Centrifugal melt electrospinning devices

Centrifugal electrospinning is a new technique for the preparation of nanofibers by the combination of electrostatic and centrifugal forces with electrospinning and centrifugal spinning. It has a dual role, centrifugal force and electrostatic force, so that the fiber is more fully stretched, and thus can produce finer fibers. Not only that, but centrifugal electrospinning is conducive to the preparation of ordered nanofibers, to improve the spinning efficiency, and be conducive to mass production. In recent years there have been many reports about centrifugal electrospinning. Li et al. [34] proposed a double spinning centrifugal electrospinning device. The influence of voltage and centrifugal velocity on the spinning fibers was investigated using a self-made spinning apparatus. Ordered fibers were obtained at voltages below 3 kV. Yan Yurong et al. [35] constructed a self-made disc centrifugal electrostatic spinning machine. Li Shijiang et al. [36] reported in their patent a centrifugal electrostatic spinning device, with the use of metal mesh made of a rotary drum to achieve centrifugal electrospinning. Liu et al. [37] prepared fluorescent nanofibers with an ordered arrangement, a cross-array, and a stranded structure, using a low-pressure centrifugal electrospinning apparatus. In the field of centrifugal electrospinning, the research group independently designed and processed a centrifugal electrostatic spinning machine, and applied for two patents [38–40].

As shown in Fig. 2.11, the main components in the centrifugal solution electrospinning device include: a rotating cylinder and an umbrella nozzle composed of spinning parts, a high-pressure generator and the collection ring with a high-voltage electric field for the jet to provide an electric field force, a frequency converter and rotating motor components (for the rotation of the cylinder to provide centrifugal force), and the support platform. The centrifugal melt electrospinning device is more complex than the centrifugal solution electrospinning device. On the basis of the above-mentioned device, the

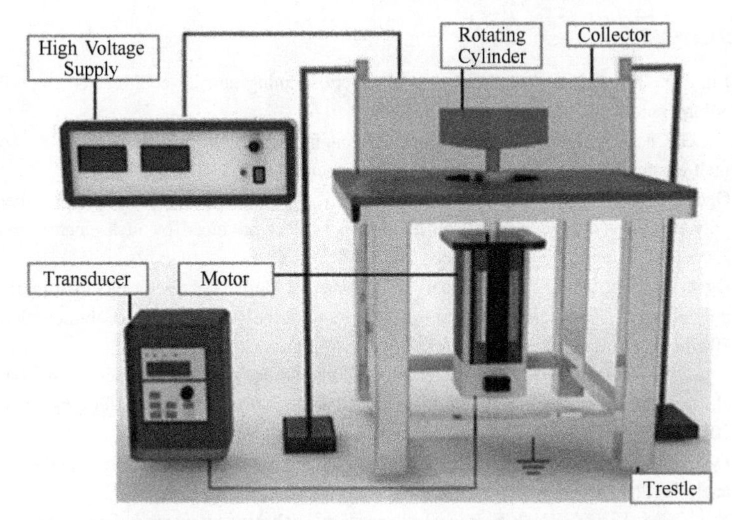

FIGURE 2.11 Device of centrifugal solution electrospinning [41].

centrifugal melt electrospinning device increases the temperature control system, which is composed of four parts: thermocouple, thermocouple bracket, temperature control box, and electromagnetic heating ring. Centrifugal electrospinning is still in the experimental stage, and most of the centrifugal electrostatic spinning devices are self-designed to overcome traditional electrospinning defects [40].

There are currently two key difficulties in centrifugal melt electrospinning devices. The first is high pressure, high temperature, and high-speed rotation. At the same time, safe operation has become one of the key issues in centrifugal melt electrospinning devices. The second point is regarding the ability of the high-speed rotating device to improve the temperature measurement. Wireless infrared temperature measurement and wired thermocouple temperature measurement are now the most commonly used.

2.10 Conclusion

Compared to the solution electrospinning device, the melt electrospinning device is not mature enough. The main reason for this is the need to apply high voltage and higher temperature, and the device is complex, it is easy to produce failure. The fiber diameter is thick, and a reduction in fiber diameter has become a key technical problem. However, in biomedical engineering, filtration, and clean energy, solvent-free melt electrospinning has more advantages than solution electrospinning, but we believe that with the continuous efforts of scientific research workers, melt electrospinning technology will continue to develop, and will be applied in many areas.

References

[1] Liu TQ. Research development on melt-electrospinning and its equipments. New Technology and New Process 2009;(12):93–6.

[2] Xie G, Chen ZY, Ramakrishna S, Liu Y. Orthogonal design preparation of phenolic fiber by melt electrospinning. Journal of Applied Polymer Science 2015;132(38).

[3] Ogata N, Lu G, Iwata T, Yamaguchi S, Nakane K, Ogihara T. Effects of ethylene content of poly(ethylene-co-vinyl alcohol) on diameter of fibers produced by melt-electrospinning. Journal of Applied Polymer Science 2007;104(2):1368–75.

[4] Ogata N, Shimada N, Yamaguchi S, Nakane K, Ogihara T. Melt-electrospinning of poly(ethylene terephthalate) and polyalirate. Journal of Applied Polymer Science 2007;105(3):1127–32.

[5] Ogata N, Yamaguchi S, Shimada N, et al. Poly(lactide) nanofibers produced by a melt-electrospinning system with a laser melting device. Journal of Applied Polymer Science 2007;104(3):1640–5.

[6] Lyons JM. Melt-electrospinning of thermoplastic polymers: an experimental and theoretical analysis (Ph.D.Thesis) . USA: Drexel University; 2004.

[7] Yang WM, Li HY, Wu WF. Recent advances in melt electrospinning. Journal of Beijing University of Chemical Technology (Natural Science) 2014;(04):1–13.

[8] Wan YQ. Research on electrospinning process behavior and vibration-electrospinning technology. Donghua University; 2006.

[9] He JH, Liu Y, Xu L. Vibrating electrospinning device for nanofiber. 2007. CN 1986913A.

[10] Chi L, Yao YY, Li RX, et al. Recent advance in manufacture of nano-fibers by electrospinning. Progress in Textile Technology 2004;5:1–6.

[11] McCann JT, Marquez M, Xia Y. Melt coaxial electrospinning: a versatile method for the encapsulation of solid materials and fabrication of phase change nanofibers. Nano Letters 2006;6(12):2868–72.

[12] Lu D, Liu TQ, Yu JX. Preparation of polystyrene hollow submicro-fiber by coaxial electrospinning. Polymer Materials Science and Engineering 2008;24(12):172–5.

[13] Wang B, Hu P. Starting from "nano spider" – global industrialization of electrospinning. New Materials Industry 2007;6:58–62.

[14] Mo XM, Wu JL, Ke QF, et al. Needleless electrospinning of polystyrene fibers with an oriented surface line texture. Journal of Nanomaterials 2013;2012:1–7.

[15] Bhattacharyya I, Molaro MC, Braatz RD, et al. Free surface electrospinning of aqueous polymer solutions from a wire electrode. Chemical Engineering Journal 2016;289:203–11.

[16] Adomaviciene M, Stanys S, Demsar A, et al. Insertion of Cu nanoparticles into a polymeric nanofibrous structure via an electrospinning technique. Fibres and Textiles in Eastern Europe 2010;18(1):17–20.

[17] Jirsak O, Santetrnik F, Lukas D, et al. Production of nanofibers from polymer solution comprises supplying polymer solution into electric field for spinning using surface of rotating charged electrode, while creating spinning surface to reach high spinning capacity. 2004. European Patent, CZ200302421-A3.

[18] Lukas D, Sarkar A, Pokorny P. Self-organization of jets in electrospinning from free liquid surface: a generalized approach. Journal of Applied Physics 2008;103(8):343–9.

[19] Yarin AL, Zussman E. Upward needleless electrospinning of multiple nanofibers. Polymer 2004;45(9):2977–80.

[20] Thoppey NM, Bochinski JR, Clarke LI, et al. Unconfined fluid electrospun into high quality nanofibers from a plate edge. Polymer 2010;51(21):4928–36.

[21] He JH, Kong HY, Yang RR, et al. Review on fiber morphology obtained by bubble electrospinning and blown bubble spinning. Thermal Science 2012;16(5):1263–79.

[22] Liu Y, Li J, Tian Y, et al. Multi-physics coupled FEM method to simulate the formation of crater-like Taylor cone in electrospinning of nanofibers. Journal of Nanoparticle Research 2014;27:153–62.

[23] Chen RX, Wan YQ, Si N, et al. Bubble rupture in bubble electrospinning. Thermal Science 2015;19(4):1141–9.

[24] Liu Y, He JH. Bubble electrospinning for mass production nanofibers. International Journal of Nonlinear Sciences and Numerical Simulation 2007;8(3):393–6.

[25] Liu Y, He JH. Control of bubble size and bubble number in bubble electrospinning. Computers and Mathematics With Applications 2012;64:1033–5.

[26] Fang J, Zhang L, Sutton D, et al. Needleless melt-electrospinning of polypropylene nanofibres. Journal of Nanomaterials 2012;2012(3):4661–70.

[27] Li M, Long YZ, Yang D, et al. Fabrication of one dimensional superfine polymer fibers by double-spinning. Journal of Materials Chemistry 2011;21(35):13159–62.

[28] Kim IG, Lee JH, Unnithan AR, et al. A comprehensive electric field analysis of cylinder-type multi-nozzle. Journal of Industrial and Engineering Chemistry 2015;31:251–6.

[29] Park J, Park JC, Jong CP. Bottom-up electrospinning apparatus for mass production of nanofibers has spinning nozzles in middle and in end with different height among spinning nozzles arranged along horizontal line. 2008. European Patent, EP1975284-A2.

[30] Lin T, Niu HT. Fiber generators in needleless electrospinning. Journal of Nanomaterials 2012;2012(3):127–40.

[31] Chitral JA, Shesha HJ. Fundamentals of electrospinning and processing technologies. Particulate Science and Technology 2016;34(1):72–82.

[32] Yan YR, Qiu ZM, Chen Q. In: Mechanism of electro-centrifugal spinning, abstract of the 17th session of the 28th annual Academic Meeting of the Chinese Chemical society; 2012.

[33] Zhang JN. Study of melt electrospinning and its mesoscope simulation in the coupling fields. Beijing University of Chemical Technology, 2017.

[34] Li SJ, Li J, Zhang YC, et al. Centrifugal electrospinning device, CN102061530A. 2011.

[35] Liu SL, Huang YY, Han YM, et al. Fabrication of fluorescent nanofibers of aligned arrays, two-layer grid-patterns and nanoropes via electrospinning 2013;26(1):44–9.

[36] Liu Y, Song TD, Chen ZY, et al. Intermittent centrifugal melt electrospinning device, CN 103215662A. 2013.

[37] Liu Y, Li XH, Chen ZY, et al. Centrifugal electrospinning device, CN 103215664A. 2013.

[38] Liu Y, Song TD, Chen ZY, et al. Intermittent centrifugal melt electrospinning device. CN 203360643A. 2013.

[39] Liu Y, Wang ZC, Xiang W, et al. Novel centrifugal melt electrospinning device. CN 203238358A. 2013.

[40] Korjenic A, Zach J, Hroudová J. The use of insulating materials based on natural fibers in combination with plant facades in building constructions. Energy and Buildings 2016;116:45–58.

[41] Peng H. Experiment and Simulation of centrifugal melt electrospinning, Beijing University of Chemical Technology, 2017.

Chapter 3

Formation of fibrous structure and influential factors in melt electrospinning

Chapter outline

Melt Electrospinning. https://doi.org/10.1016/B978-0-12-816220-0.00003-8

There are many factors that affect fiber diameter and distribution during the melt electrospinning process, which usually include the molecular structure of the raw material, the spinning temperature, the voltage, the spinning distance, and the ambient humidity. All factors synergize, and the impact is different. At present our research group has utilized various polymer samples such as polypropylene (PP), polyethylene (PE), polyamide (PA), polyethylene terephthalate (PET), polylactic acid (PLA), and polycaprolactone (PCL), and so on to manufacture fiber by melt electrospinning. The effects of voltage, temperature, receiving distance, and electric field distribution on the morphology and properties of the fibers have been studied, and the microscale fibers have been successfully prepared. This chapter mainly introduces the experimental process of typical materials made in our laboratory, as well as a discussion of the experimental results. Finally, the development prospects and trends of melt electrospinning are summarized in Section 3.5.

3.1 Polycaprolactone

Poly (ε-caprolactone) (PCL), also known as polycaprolactone, is a synthetic biodegradable polyester, which has attracted attention in recent years, in various fields, especially in biomedial fields. PCL is a kind of thermoplastic polyester prepared by the ring-opening polymerization of ε-caprolactone. The glass transition temperature is −60°C, the melting temperature is 63°C, and decomposition starts at 250°C. PCL has good thermal stability, hydrolytic stability, and low-temperature characteristics. In addition, PCL has a longer degradation time than PLA, which makes it a promising material for applications requiring longer degradation time. PCL is a biodegradable and nontoxic polymer material that can be completely decomposed into CO_2 and H_2O in soil and water environments. In addition, PCL has good drug passability and biocompatibility, and can release drugs for a long time. The material is currently approved by the US Food and Drug Administration (FDA) and is widely used in fracture fixation materials, surgical sutures, medical dressings, drug-controlled release materials, tissue engineering scaffold materials, and other fields. PCL has rigidity and good flexibility, but also has good high-temperature and low-temperature performance, excellent mechanical properties, and an excellent processing performance [1].

PCL fiber is a kind of absorbent medical synthetic fiber, which has good drug permeability, is widely used as a drug release carrier, for surgical sutures, etc. PCL is commonly blended with starch, which results in a good biodegradable material at a low cost. PCL mixed with other polymers enhances its properties to a wide range of applications such as biodegradable bottles, films, nonwoven fabrics, etc. At present, PCL fibers are prepared by solution electrospinning. Generally, to prepare a 12% (W/V) spinning solution, PCL raw materials are dissolved in 50:50 (V/V) DCM/DMF solutions, and then the fibers are spun at a high voltage of 21 kV [2]. However, the solution of the DMF or acetone used in the electrospinning solution is toxic [3], and the spinning efficiency is very low, generally 0.001 $g \cdot h^{-1}$.

3.1.1 Experiment

In order to overcome these shortcomings, our group explored the melt electrospinning PCL fiber [4]. The experimental setup used in this study is the high-efficiency melt electrospinning device [5] shown in Fig. 3.1, and the double-point nozzle is used [6]. The materials employed in the experiment have an average molecular weight of 40,000 $g \cdot mol^{-1}$, the melting point is 59−64°C, and the glass transition temperature is −60°C.

3.1.2 Results and discussion

We designed and built the patent melt electrospinning apparatus shown in Fig. 3.1. During the experimental process the viscosity of the PCL melt was

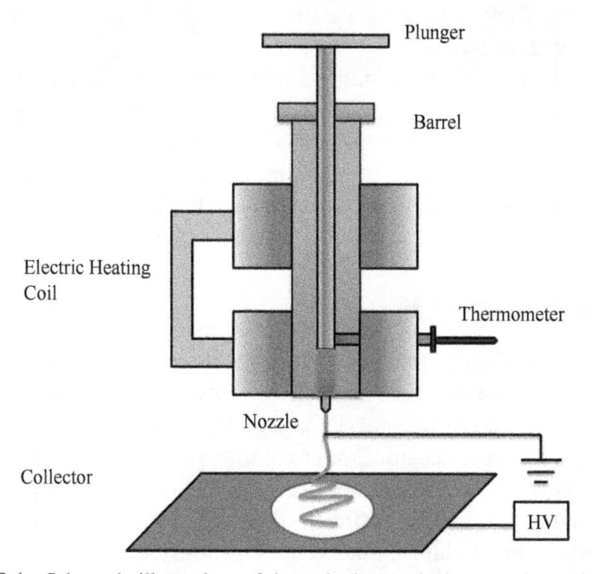

FIGURE 3.1 Schematic illustrations of the melt electrospinning experimental apparatus [5].

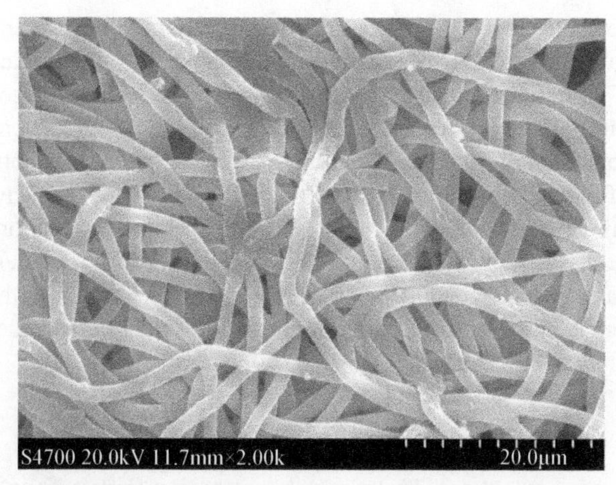

FIGURE 3.2 SEM images of PCL melt electrospun fibers [4].

large, and it was spinning at a decomposition temperature of 200°C. Through repeated tests, the ultrafine PCL fiber was successfully spun when the temperature of the cylinder was raised to 240°C and pure PCL fibers were obtained directly, in which there were no poisonous materials. The average diameter of the fibers was 1.35 μm and the spinning efficiency was about 5 $g \cdot h^{-1}$,which was much higher than with solution electrospinning. It can be seen from Fig. 3.2 that due to the high spinning temperature (close to its decomposition temperature), there may be a small number of PCL molecules which are decomposed during the spinning process. This situation, caused by the fiber curing, is not sufficient to produce the viscosity of the larger part of the phenomenon of fiber bonding.

Through this melt electrospinning experiment for PCL, on the one hand, there is a need to improve the flow of PCL raw materials, so that it can be less than the decomposition temperature of the smooth spinning. On the other hand, it is necessary to improve the cooling conditions after spinning, so that the fiber surface temperature is rapidly reduced. For example when the fibers are received in cold water, a non-tacky fibrous nonwoven fabric is obtained.

3.2 Polylactic acid (PLA)

PLA is one of the most promising biodegradable polymers and has attracted attention over past decades. It has superior properties to those of PE, PP, polystyrene (PS), polycarbonate (PC), and PET. L-Lactic acid provides a polymer with a higher mechanical strength than the DL-form [7], particularly in terms of tensile strength [8], as the pure L-form has a high degree of crystallinity, and thus there is sufficient mechanical strength only in its fully isotactic form. In some chitosan-based composites, the addition of PLA leads to an improvement in the modulus [9]. Moreover, the degradation products of polylactides are nontoxic, which enhances practical applications in biomedicine and other fields [10,11]. PLA is well known as a biodegradable polymer in daily applications, such as disposable cutlery, plates, cups, lids, packaging, films, and containers for liquid foods [12]. In most applications of biodegradable polymers and composites, PLA is one of the important candidates for high-value side and specific applications, such as tissue engineering, prosthetic devices, implants, catheters, sutures, and anticancer drug delivery [11]. In addition to biomedical usages, the employment of the biodegradability of PLA in packaging [13] and agricultural materials is worth highlighting, because PLA and its fibers can eventually be decomposed into carbon dioxide and water through the actions of microorganism and, finally, these degraded items become recyclable materials [14,15]. Due to the biodegradability and recyclability of PLA in nature, PLA has been referred to as the environmental cycle material of the 21st century [16]. Many investigations have recently been conducted on PLA fibers [17–19]. Although having these kinds of advantages, PLA has some drawbacks, which limit its application in certain fields. For instance, PLA is not utilized in the biomedical field due to its slow biodegradable rate and hydrophobicity.

PLA fibers are generally produced industrially by solution and melting processes, but the resulting fibers are thicker. In recent years, in addition to the traditional spinning process, the longstanding fiber preparation method of electrospinning has aroused great attention from researchers and has become a hot research topic, both at home and abroad. This is because the electrospinning of polylactic acid ultrafine fiber in tissue engineering scaffold materials, wound covers, new drug-release carriers, and other biomedical fields has broad application prospects. At present, people often use the solution of electrospinning PLA superfine fiber. Generally, the PLA raw material is dissolved in dichloromethane or chloroform to form a spinning solution of 8%–16% (W/V), and then the fibers are prepared at a direct current (DC) high voltage of 21 kV [20–22]. However, the solvent used for the electrospinning of dichloromethane or chloroform is toxic and requires special treatment to be applied, greatly increasing the production cost of PLA fiber. In addition, its spinning efficiency is very low, generally 0.001 $g \cdot h^{-1}$. So far, the use of solution spinning PLA fiber is still in the experimental stage, and it has not yet been seen in commercial production reports [23], therefore countries focus on research and development on melt spinning. However, since the viscosity of the polymer melt is much greater, relative to the solution, a greater electric field force is required. But the electric field force cannot infinitely increase, and therefore, nano-grade fiber cannot be successfully prepared by the melt electrospinning method. Therefore, our laboratory, in addition to studying the performance of PLA fiber, has also explored the low-cost, high-efficiency, green preparation of PLA superfine fiber.

In order to overcome these shortcomings, we explored the melt electrospinning method to prepare PLA ultrafine fiber and its performance exploitation. In 2010, Liu Y et al., studied the melt electrospinning process for PLA [24]. In the experiment, the high-efficiency melt electrospinning device [5] was designed and used, and a double-sprinkler was selected [25,26], as shown in Fig. 3.3. The experimental conditions for the preparation of polylactic acid ultrafine fibers were explored. The raw materials used in the experiment were PLA (grades 2002D) from Nature Works LLC, melting at 210°C with a melt flow rate of $8\ g\cdot(10\ min)^{-1}$. The raw materials were dried beforehand (70°C, 4 h). The prepared PLA fibers were observed by scanning electron microscopy (SEM), as shown in Fig. 3.4. The spinning conditions were spinning voltage 60 kV, receiving distance 11 cm. We fabricated PLA fiber at two different temperatures to study the effect of increasing temperature. Figs. 3.4A and B are SEM images of PLA fiber electrospun at 230 and 245°C, respectively. By comparing the two results, we found that PLA material electrospun at 230°C produced ultrafine fiber with an average diameter of 1.6μm, fiber diameter variance was 0.21μm,and spinning efficiency was about $4\ g\cdot h^{-1}$. Even though the

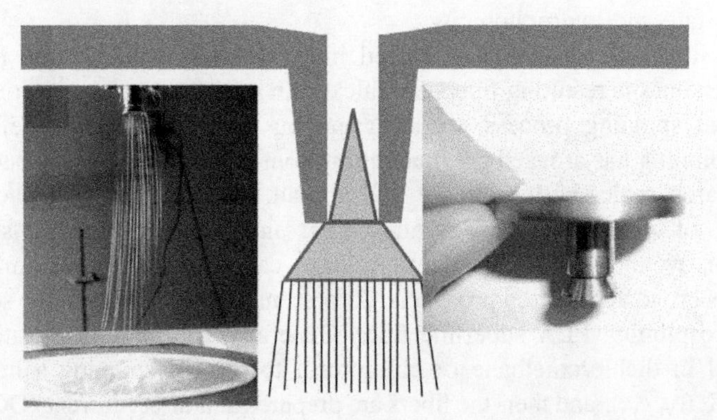

FIGURE 3.3 High-efficiency melt electrospinning spinning process [25, 26].

FIGURE 3.4 SEMs of (A) PLA fibers spun at 230°C; (B) PLA fibers spun at 245°C [24].

spinning temperature (230°C) was a little higher than the melting temperature of PLA material, it exhibited a uniform fiber diameter and smooth surface. Meanwhile, the fibers electrospun at 245°C exhibited a nonuniform fiber diameter and were broken even though the same spinning conditions were maintained for 230°C. During the spinning process, it was found that the fibers which collected on the collection device were continuous and not broken, but the collected fibers were brittle and easily twisted into powder. Fig. 3.4 shows a fiber topography at a spinning temperature distribution of 230°C and 245°C. It can be seen from the figure that the fiber is continuous at 230°C, and the fiber morphology is broken at a temperature of 245°C. From the above analysis we concluded that PLA melt electrospinning should try to avoid high-temperature processing (below 230°C is appropriate), so as not to affect the mechanical properties of the PLA fiber.

In order to alleviate the thermal decomposition and obtain superfine fibers at low temperature, Zhao et al. [27] tried to add sucrose fatty acid esters (SE) to decrease the viscosity of the PLA melt. In the present study, SE was added to increase the flow ability of PLA melt, which ensured PLA fibers of smaller diameters were produced at low temperature. The low electrospinning temperature decreased the degradation of PLA.

In 20 g of PLA, 8% of SE was added using a miniature extruder (DYNISCO, LME-230, USA). The screw temperature was 185°C. The die head temperature was 180°C. The screw speed was 30 $r \cdot min^{-1}$.

PLA melt has been electrospun at various temperature ranges of about 180, 190, and 210°C, and also with various electrospinning loadings. We fabricated the fiber with pure PLA fibers and PLA fibers with added SE, to study the effect of plasticizer and load before and after the addition. From the result obtained, due to the high viscosity of PLA melt, ultrafine fibers could not be fabricated for pure PLA at 180 and 190°C. But at 210°C, ultrafine fibers were produced for pure PLA with an average diameter of 6.36 μm. To further reduce the diameter, SE was added and electrospun at 180°C with an average diameter of 3.06 μm and 5 × 700 g load. As a low-molecular-weight material and a surface active agent, SE decreased the PLA melt viscosity, which results in a decrease in diameter. To achieve a further reduction in diameter, when electrospinning occurred at 190°C and the load was 5 × 700g , 3 × 700g, and 1 × 700 g, the average diameter of PLA fibers, in turn, was 1.89, 1.70, and 1.54 μm. Therefore, decreasing the diameter, along with decreasing the electrospinning load, produced a large quantity of PLA melt flow forming thick fiber at the same length.

In 2016, Xie et al. [28] applied a pulsed electric field to overcome the difficulty of reducing fiber diameters during melt electrospinning. The effect of the frequency and duty cycle of the pulsed electric field on the fiber diameter, crystallinity, and molecular orientation was studied. The results revealed that the diameter of the poly-L-(lactic acid) (PLLA) melt electrospun fiber was reduced by this pulsed electric field, especially at higher frequencies, and the finest fiber was obtained at a duty cycle of 29.8% and a frequency of 1 kHz. Both constant and pulsed electric fields can lead to a high molecular orientation in electrospun fibers, particularly in the crystalline regions. The specific experiments are as follows.

3.2.1 The diameter of PLLA fiber under a pulsed electric field

PLLA particles ($M_n = 100{,}000$; Zhejiang Hisun Biomaterials Co. Ltd., China) were dried at 60°C for 4 h in a vacuum oven before the experiments began. Fiber samples were prepared with the same amount of fresh PLLA particles. The melt electrospinning devices used were the same as those in our previous studies [29,30]. However, in this study we utilized a square wave pulse power supply (DMC-200, Dalian Ding Tong Technology Co. Ltd., China) with adjustable voltage, frequency, and duty cycle. The collector, covered with aluminum foil, was connected to the positive side of the power, and the spinneret was grounded to avoid interference between the high voltage and the thermal system. Accordingly, the electrospun process was stable, and no influential thermal decomposition occurred when the cylinder temperature was 240°C. The spinning distance was 10 cm, and the voltage was 40 kV. Although the flow rate relied on the self-gravity of the melt and the mass was tiny, the flow rate was fixed. A schematic diagram of the experimental procedure is illustrated in Fig. 3.5A. Fig. 3.5B shows the applied pulse voltage waveform. Samples were also prepared using a normal power supply [DW-P503-2ACDE; Dong Wen high-voltage power supply (Tianjin) Co. Ltd., China] for comparison.

3.2.1.1 Characterization and measurement

The morphology of the obtained fibers was characterized using SEM (Hitachi S4700, Japan) at an accelerating voltage of 20 kV, and each sample was

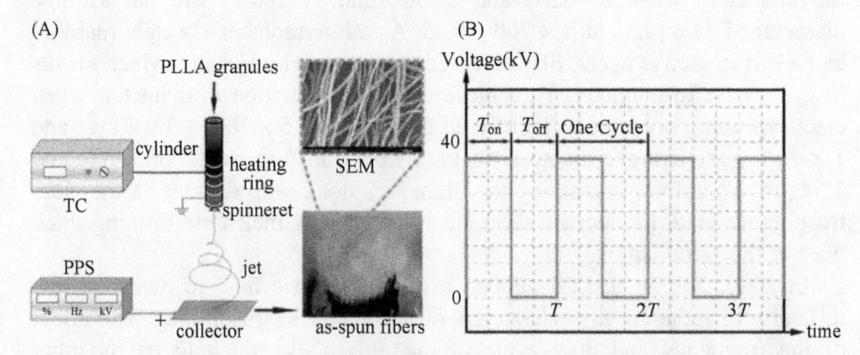

FIGURE 3.5 (A) Schematic diagram of the experimental procedure: TC represents the temperature controller, and PPS represents the pulsed power supply. (B) Schematic diagram of the output voltage waveform signal (*blue straight line*) of the pulse power supply. T represents the cycle. Frequency is equal to $1/(T_{on} + T_{off})$. The duty cycle is equal to T_{on}/T_{off} [28].

coated with gold before imaging. The average fiber diameter and its standard deviation were calculated from the SEM images by using Image Pro Plus 6.0 software. There were 15 fibers randomly selected from the SEM image of each sample, and each fiber was measured at five different locations.

3.2.1.2 Results and discussion

The electrospinning process can be affected by a variety of parameters, including material conditions, viscosity, or conductivity, and processing variables such as applied voltage, spinning distance, flow rate, and temperature [31]. In this contribution, the influence of frequency and the duty cycle of the pulsed electric field, during the melt electrospinning process, is primarily discussed. Therefore, the other conditions are all kept fixed. In the early exploration experiments, we found that the corona phenomenon occured frequently when the frequency exceeded 5 kHz, and although this may be related to environmental conditions, we first carried out experiments at a duty cycle of 49.8%, to explore the influence of frequency (1, 1.49, 2.5, 4.46, and 4.95 kHz) on the fiber diameter. For comparison, samples were also prepared using normal electrospinning under the same conditions.

The relationship between frequency and fiber diameter was explored. As can be seen from Fig. 3.6, the average fiber diameter decreased as the frequency increased; all the diameters were smaller than the corresponding samples made via normal electrospinning. At the same time, the distributions of diameters in the pulsed electric field were larger than those in the constant electric field. The square wave pulse voltage provided an intermittent electric field, different from the constant electric field. This caused the action time,

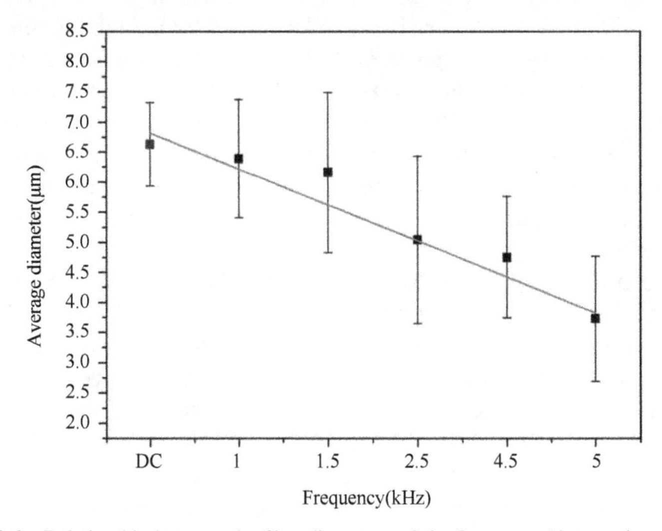

FIGURE 3.6 Relationship between the fiber diameter and the frequency (duty cycle = 49.8%) [28].

when using the pulsed electric field, to be about half that of the constant electric field, which resulted in coarser fibers. However, in one cycle of the pulsed electric field, the jet decelerated due to inertia during the half cycle as the voltage was reduced to zero. Simultaneously, the normally tense flexible molecular chains relaxed and contracted in the elastic range during this period. During the half cycle with voltage at 40 kV, the entangled molecular chains tensioned and elongated due to the electric force present. In the half cycle with no voltage, the chain relaxation may have caused the entangled molecular chains to become partially disentangled. Under the intermittent and repeated impact of the pulsed electric force, disentanglement of the molecular chains continues and possibly occurs at a faster rate, which promotes molecular orientation and jet slenderizing and reduces the final fiber diameter. Richard et al. [32] also thought that partial disentanglement would explain decreasing fiber diameters. On the other hand, since the amplitude of the pulse voltage is the same, the duty cycle is also consistent, and high-frequency voltage has more pulses over a given time period than low-frequency voltage. During this time, the total action times of the different pulse voltages are equal. In other words, the only difference between high frequency and low frequency is that the change of the intermittent voltage is shorter and faster under high-frequency conditions. This may cause the jet to accumulate more energy, imparting a greater kinetic energy for stretching. Therefore, the fiber diameter decreases with increasing frequency. Overall, we realized that the pulsed form of energy accumulation and the chain disentanglement effects provided by pulsed electric fields were the reasons for obtaining smaller fiber diameters under the selected frequencies. In addition, the effect of a pulsed electric field on the liquid surface and Taylor cone was similar to the dielectrophoresis force proposed by Ghashghaie et al. [33]. Prior to the formation of the jet, the pulsed electric field provided an intermittent and repeated force to the liquid surface and Taylor cone, so that dipoles of PLLA rotated in the Taylor cone and aligned along the electric field lines, which facilitated the jet formation and stretching, in agreement with previous reports [29,34]. Regarding the fiber uniformity, the diameter distribution of fibers produced in a pulsed electric field was broader than those in a constant electric field, which can be attributed to the presence of the acceleration and deceleration of the jet in the pulsed electric field.

The relationship between fiber diameter and duty cycle at a frequency of 1 kHz was also investigated, for duty cycles of 1.2%, 9.3%, 29.8%, 39.7%, and 49.8%. Surprisingly, fibers were obtained even at a duty cycle of 1.2% (Fig. 3.7), and the fibers were thinner than those produced via normal electrospinning. Fig. 3.7 presents a concave curve showing the relationship between the fiber diameter and duty cycle. In general, the electric field action time becomes longer when the duty cycle is increased, resulting in greater jet stretching and finer fibers. However, the average diameter increases when the duty cycle increases past 29.8%. This suggests that there may exist an optimal

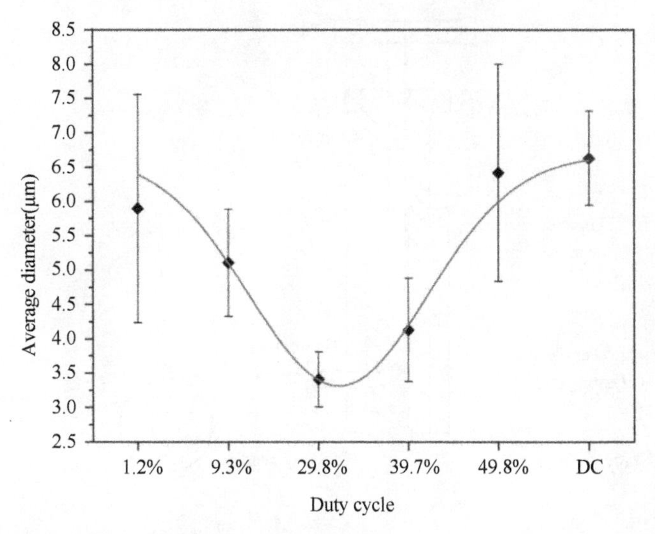

FIGURE 3.7 Relationship between the fiber diameter and the duty cycle (frequency = 1 kHz) [28].

duty cycle for a given frequency to achieve the finest fibers. Under these conditions, the electric force and disentanglement provide the greatest contribution to the refinement of the fibers.

3.2.2 Thermal degradation of PLA fiber

Electrostatic electrospinning is a simple and safe method for producing ultrafine fibers compared to solution electrospinning. However, higher spinning temperatures typically result in severe degradation of the polymer material. In view of the above-mentioned thermal degradation problems, our research group [27,35,36] used self-made spinning nozzle equipment to produce high-efficiency fibers, by studying the spinning temperature, spinning distance, and the type of additives to reduce the thermal degradation of PLA. The specific experimental content is as follows.

3.2.2.1 Materials

L-PLA (melting point: 210°C) was purchased from Nature Works LLC, USA (trademark: 2002D). The PLA sample was baked at 70°C for about 4 h before spinning. Antioxidants 1010 and 168 were purchased from Beijing Additive Institute.

3.2.2.2 Equipment

A diagram of the melt electrospinning equipment [25,35], specifically self-designed for this study, is shown in Fig. 3.8. The cylinder, piston, and spray head were made of steel. The heating system included electrical heating rings,

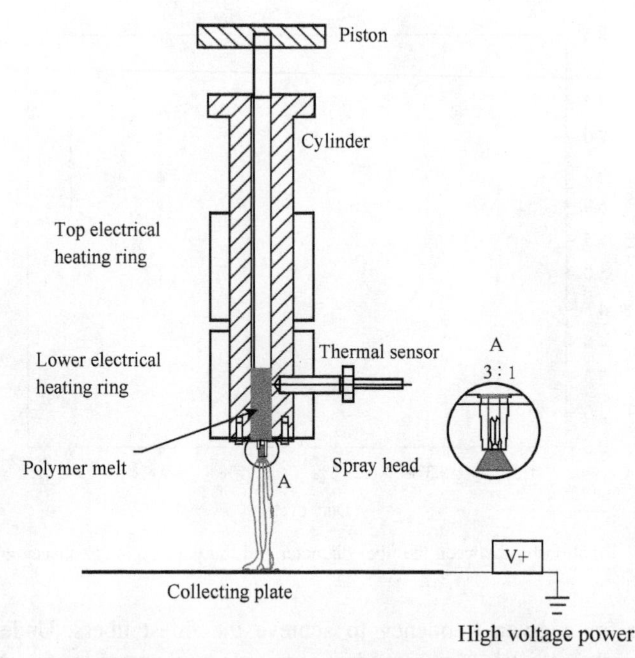

FIGURE 3.8 Scheme of the melt electrospinning apparatus [36].

thermal sensors, and a temperature control subsystem. The spray head was a cone-type instead of the traditional capillary tube. The melts the surface of an umbrella-like spray head and then formed Taylor's cones automatically at the bottom edge of the head in the high-voltage electrostatic field [35].

3.2.2.3 Preparation of PLA fibers

The cylinder was heated to the required temperature (200–250°C). Then, the granular or powdery material was put into the cylinder at the preset temperature for approximately 10 min (i.e., until the melts were spread uniformly on the spray head), the high-voltage supply device was switched on and adjusted to an appropriate value (30–100 kV). Thus, the melts formed several cones along the bottom edge of the spray head, and several strips of the PLA melt flew to the collecting mesh, as shown in Fig. 3.9. The distance between the collecting mesh and spray head was set at around 10–23 cm.

3.2.2.4 Characterization

The morphologies of the electrospun fibers were observed by SEM (Hitachi S4700). Fiber samples were coated with a 10nm layer of platinum before the observation. The scanning voltage was 20 kV. The relative molecular mass of the PLA fibers was measured by a gel permeation chromatograph (Shimadzu LC-6A). X-ray diffraction (XRD) patterns of the fibers were recorded using a

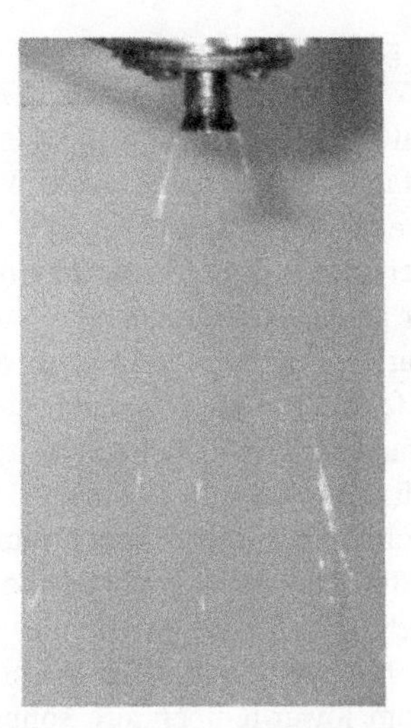

FIGURE 3.9 Melt electrospinning with an umbrella-like spray head [36].

Rigaku D/Max-IIIC (CuKα radiation). The thermal transition curves of the PLA samples were recorded using differential scanning calorimetry (DSC; Perkin Elmer Pyrisl) to demonstrate the crystalline state of the PLA before and after electrospinning.

It is well known that temperature has significant effects on melt electrospinning [25,37]. Overheating at high temperatures leads to a sharp decrease in the melt viscosity, indicating a decrease in molecular weight. Consequently, the polymer melts flow faster and the residence time in the spinning device is shorter. An SEM micrograph of the PLA fibers obtained from melt electrospinning with the cylinder temperature set at 210°C and at a voltage of 60 kV is shown in Fig. 3.10A. The average diameter of the resultant fibers was 7.65 μm, with a fiber diameter variance of 0.21 μm and the spinning efficiency

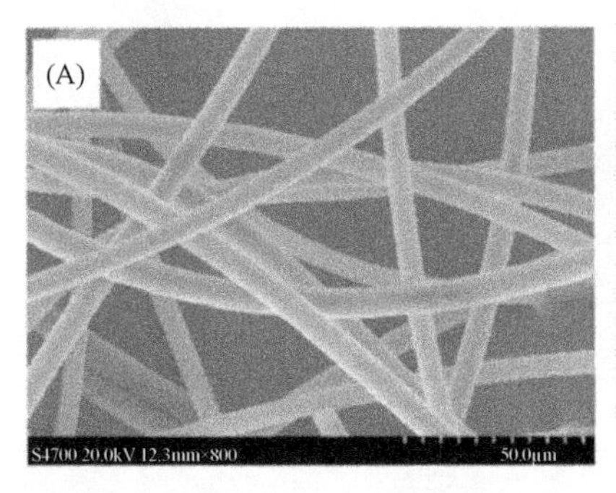

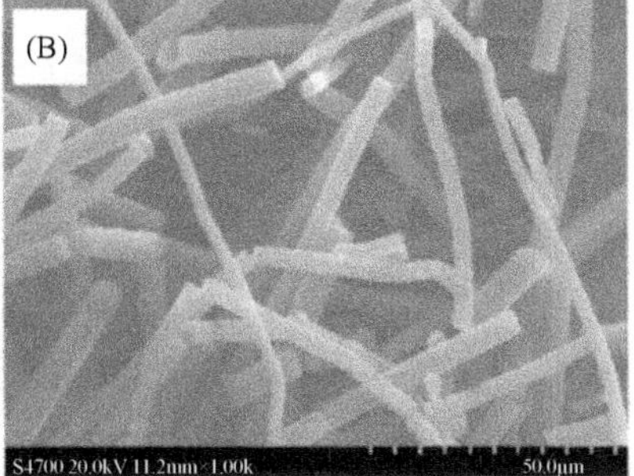

FIGURE 3.10 SEM of (A) PLA fibers spun at 210°C, (B) PLA fibers spun at 245°C. The distance between the spray head and collection mesh was 11 cm [36].

of electrospinning was 4 g·h^{-1} when using the abovementioned conditions. It was noticed in Fig. 3.10A that the surface of the PLA fiber was quite smooth, suggesting the fibers might have higher tensile strength compared with those with defects on the surface; such defects result in typical poor mechanical properties from solution electrospinning.

However, if the temperature is higher than the limit, the polymer will start to degrade, especially for an easily degradable polymer. From a DSC scan, it was evident that PLA start to degrade significantly at 250°C. An SEM image shows that PLA fibers fabricated at a spinning temperature of 245°C in Fig. 3.10B. Apparently, many of the PLA fibers were broken into short fibers and the fiber diameter distribution was broader. However, during spinning, fibers were seen on the collection mesh with continuity and were unbroken. It could be concluded that the short fibers might be due to the pressing force from the tweezers when the electron microscopy samples were prepared. Although such an augmentation might be reasonable, the pressing force was also applied to the fibers resulting from lower-temperature spinning (e.g., 210°C). Thus, temperature is a key parameter necessary to be taken into consideration and to be well controlled during PLA melt spinning.

Heating is a necessary condition for melt electrospinning for us to know the extent of degradation of the spun fibers. To determine the intrinsic relationship between thermal spinning at a certain temperature and the extent of degradation, the relative molecular mass (M_p, as a function of molecular mass vs. eluent volume) of PLA before and after melt electrospinning at 245°C was measured. A comparison of the M_p of the PLA before and after melt electrospinning at 245°C is shown in Fig. 3.11 and given in Table 3.1. The M_p of the PLA decreased (by 56.2%) after melt electrospinning. The possible reason for this is that high spinning temperatures (e.g., above 230°C)

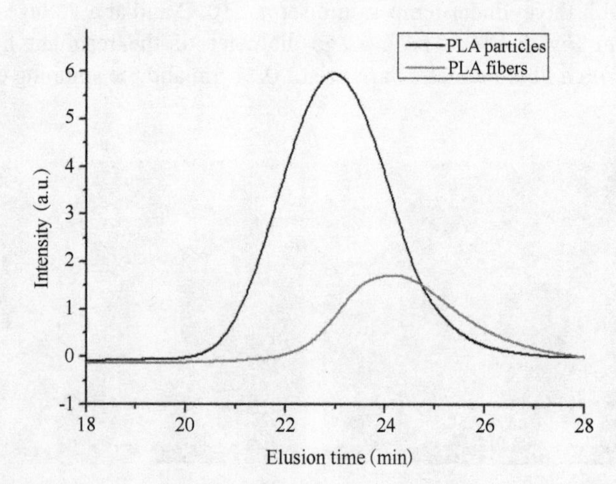

FIGURE 3.11 Comparison of relative molecular mass of PLA before (articles) and after (fibers) melt electrospinning at 245°C [36].

TABLE 3.1 Relative molecular mass of PLA before and after melt electrospinning [36] **.**

PLA	M_p	M_n	M_w/M_n
Particles	131,277	92,736	1.8
Fibers	57,497	34,733	2.0

cause thermal degradation of the PLA molecules. It was reported from some available experimental data, that at temperatures above 200°C, PLA may experience a minimum of four reaction pathways during thermal decomposition:

(1) Intramolecular and intermolecular ester exchange to produce lactide and cyclic oligomers;
(2) cis-Elimination to generate acrylic acid and acyclic oligomers;
(3) Radical and concerted nonradical reactions to produce CH_3CHO and CO;
(4) Radical reactions to generate $CH_3CH{=}C{=}O$, CH_3CHO, and CO_2.

Among these possible pathways, there are two principal pyrolysis mechanisms for polyesters: cis-elimination and transesterification. But at temperatures above 300°C, nonselective radical reactions may also occur. Therefore, the outcome in Fig. 3.11 and Table 3.1, for PLA breaking down into short-chain molecules, can be readily understood. It can be concluded that melt electrospinning for pure PLA material will naturally lead to a certain extent of degradation of PLA at above 200°C because heating is necessary for PLA melting and then for the spinning process.

Degradation leads to a decrease in tensile strength because PLA is either a crystalline or a partially crystalline polymer, and tensile strength is relevant to its isotactic form. The tensile strength of individual fibers is difficult to measure because the fibers are very fine and thin. The degree of crystallinity can reflect the strength in an aspect. Hence, the XRD and DSC patterns of PLA were recorded, as shown in Figs. 3.12 and 3.13. From Fig. 3.12, it can be seen that the PLA particles have a clear crystalline peak before electrospinning. After spinning, however, the sharp crystalline peak has disappeared and a smooth and broad amorphous band peaked at round $2\theta = 17^\circ$. The raw PLA from the polymerization reaction of L-lactic acid can form a high degree or partial isotactic structure of polymer and the isotactic structure constructs a lattice during polymerization. Consequently, a sharp crystalline peak existed in the XRD scan of the raw PLA. Comparatively, after electrospinning, the relative molecular mass of PLA became lower due to the thermal decomposition of the molecular chains. Although the chains underwent strong orientation, they could not crystallize as effectively, because the chains were short

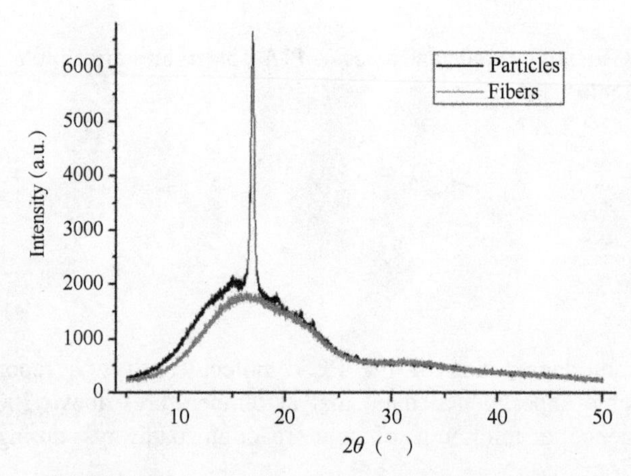

FIGURE 3.12 XRD patterns of PLA particles and fibers [36].

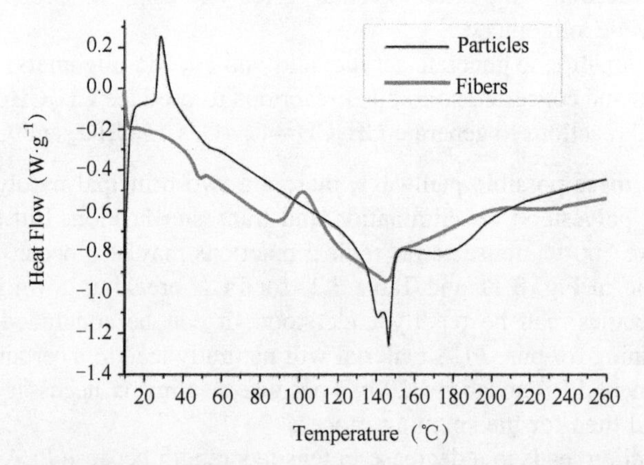

FIGURE 3.13 DSC thermograms of PLA particles and fibers [36].

and the decomposition impurities generated amorphous domains, which caused high defects in the existing lattices. The DSC thermograms of PLA before and after spinning are shown in Fig. 3.13. It can be seen that the PLA particles have two sharp melting peaks at 140 and 146°C. However, after spinning, the sharp melting peak has disappeared. The small and smooth melting peak indicated an incomplete crystalline state. Although crystallization occurred, the amount of crystallinity was too small to be registered in the XRD patterns.

3.2.2.5 *Methods in alleviating the problematic degradation of PLA spun fibers*

In order to minimize the side effects of thermal decomposition or to alleviate the thermal degradation for PLA in the melt electrospinning process, the initial electrospinning process was modified in terms of the addition of antioxidants, the equipment, and the processing conditions.

For the additives, antioxidant 1010 and antioxidant 168 were added to reduce the chain scission reaction. The antioxidant 1010 is a hindered phenolic antioxidant, which donates a hydrogen atom to form a stable aryl radical. This radical captures chain-free radicals generated from the thermo-oxidative degradation of PLA, thus partially terminating reactions at the chain break. Antioxidant 168 is a phosphite ester antioxidant. It is able to trap peroxyl radicals and effectively decompose hydroperoxides produced during the processing of polymeric materials, and hence inhibits the PLA from undergoing thermal degradation.

In terms of equipment, a low- and a high-temperature electrical heating ring, rather than two high-temperature heating rings, were installed and used to reduce the high-temperature shock/input for avoiding unnecessary time taken for the PLA feed in the cylinder. The gap between the spinning head and the cylinder was increased to ensure that the viscous melt could flow down continuously and constantly along the conical surface of the spray head, thereby decreasing the residence time of the PLA in the hot cylinder.

As for the processing conditions of electrospinning, the spinning temperature was reduced from the usual 230°C down to 210°C. A new high-voltage supply device with a maximum output of 100 kV was used instead of 60 kV. The spinning voltage can be increased from 60 to 100 kV.

3.2.2.6 Effect of additives on degradation of PLA

Antioxidants as additives can alleviate the problematic degradation of PLA fibers. After adjusting the heating temperature and adopting a new electrospinning device, a considerably higher efficiency (about 6 $g \cdot h^{-1}$) of fiber formation was obtained, as opposed to 4 $g \cdot h^{-1}$ using the initial device settings. The detailed effects of temperature and voltage on PLA fiber properties were reported previously [25]. In the present study, the impact of antioxidants on PLA degradation was examined in detail (shown in Fig. 3.14 and Table 3.2). The mass ratio of the antioxidant and PLA in the feed was 3:1000. In the case of a mixture of antioxidants, the mass ratio of antioxidant 168 to antioxidant 1010 was 1:1. The spinning voltage was set at 100 kV with a spinning distance of 12 cm. On comparing curves (a) and (b) in Fig. 3.14, it could be seen that antioxidant 168 had little effect in moderating the degradation of PLA fibers. A comparison of curves (a) and (c) indicates that the 1:1 mixture of antioxidant 168 and antioxidant 1010 had enhanced effects in alleviating degradation. In the case of the mixed antioxidants, the relative molecular mass (curve(c)) was 18.05% higher than that of the antioxidant-free counterpart (curve a). The antioxidant 1010 showed the most distinct effect in hindering the degradation of PLA fibers among the cases shown by curves (b), (c), and (d). The M_p of the PLA fibers processed in the presence of antioxidant 1010 (curve(d)) was 35.5% higher than that of the fibers fabricated in the absence of an antioxidant (curve(a)).

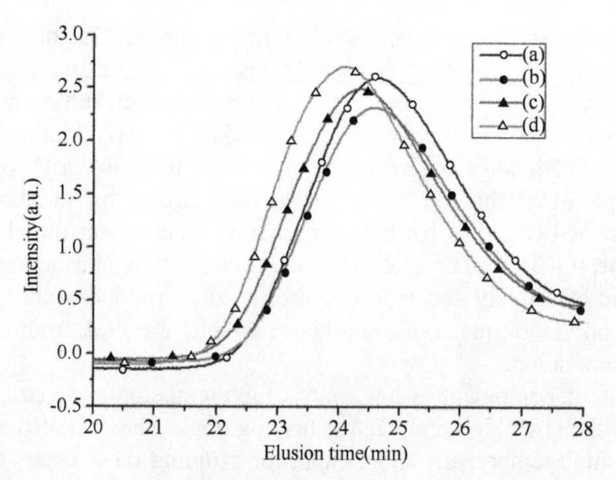

FIGURE 3.14 Comparison of the relative molecular mass of PLA fibers spun under the same temperature and processing parameters: (a) PLA fibers; (b) PLA fibers + antioxidant 168; (c) PLA fibers + antioxidant 168 + antioxidant 1010; (d) PLA fibers + antioxidant 1010 [36].

TABLE 3.2 Relative molecular mass of PLA fibers electrospun under the same temperature and processing parameters [36].

PLA	M_p	M_n	M_w/M_n
PLA fibers	35,367	7970	4.2
PLA fibers + antioxidant 168	36,919	7403	6.0
PLA fibers + antioxidant 168 and 1010	41,751	6437	4.5
PLA fibers + antioxidant 1010	48,482	9664	4.9

Finally, a comparison of Figs. 3.11 and 3.14 (a) shows that the PLA fibers produced in this study in the absence of an antioxidant have a relatively lower molecular mass than those obtained in a previous study. The PLA used in the present study can be degraded under natural conditions.

3.2.2.7 Session conclusion

The problematic degradation of facile degradable PLA in melt electrospinning was alleviated by means of hindering thermal degradation of the spun PLA fibers by the addition of antioxidants and modification of the equipment and processing parameters, particularly temperature in the study. Lowering the temperature and the addition of an antioxidant were found to be effective in alleviating the degradation of PLA in melt electrospinning. The effects of two antioxidants, individually and as a mixture, were compared and it was found that antioxidant 1010 had a remarkable capability to hinder the thermal degradation of PLA during melt electrospinning.

3.2.3 The relative molecular mass of PLA fibers

Orthogonal design [38–40] is a mathematical method used for planning multifactor tests. It is a balanced arrangement of pairs or groups and is applied broadly in many fields to optimize test designs. In this study, spinning temperature, spinning distance, and species and content of the antioxidant were selected as targets for investigation. All other factors, such as electrospinning pressure, electrospinning voltage, and ambient temperature, were maintained. Each of the four above-mentioned factors could be changed at three levels. The $L_9\ (3)^4$ orthogonal array was used to arrange the tests. The number "4" stands for four factors and "3" stands for three levels. If every factor and level was considered, there could be 81 experiments. However, factors and levels were collocated evenly, so there were only nine experiments. Balanced collocation ensured that three levels of one factor appeared three times each, and that the various levels' collocations, between two factors appearing to be the same, happened only once. Effects of the four factors on relative molecular mass were then examined. Details of the three levels of each factor are shown in Table 3.3.

Nine experiments, in accordance with the $L_9\ (3)^4$ orthogonal array, were performed. The results of the effects of the four factors on the relative molecular mass of PLA fibers are presented in Table 3.4, where M_{1j} represents the sum of the results of row "j" if the level is 1. For instance, M_{11} is the sum of the results of the first row (temperature) when the level is 1 (26,349 + 7922 + 8856 = 14,376). R_j represents the difference between the largest and the smallest M_{ij} values of row "j." For example, 9879 is the difference between 16,694 and 6815. A large value of R_j indicates that the effect of factor j on the molecule weight of PLA is significant. Based on the R_j value in Table 3.4, 11,168 > 10,001 > 9879 > 1384. Thus, the effect of the four factors on the relative molecular mass of PLA fibers can consequently be listed in the following order:

content of antioxidant > distance > temperature > species of antioxidant.

TABLE 3.3 Contents of orthogonal factor and their levels [35].

	A	*B*	*C*	*D*
Factor level	Temperature (°C)	Distance (cm)	Species of antioxidant	Content of antioxidant (%(*w*))
1	200	12	*X*	0.1
2	210	14	*Y*	0.3
3	220	16	*X*:*Y* = 1:1	0.5

X is antioxidant 1010; *Y* is antioxidant 168.

TABLE 3.4 Effects of the four factors on relative molecular mass of melt electrospinning fibers [35] .

	A	*B*	*C*	*D*	
Experiment	Temperature (°C)	Distance (cm)	Species of antioxidant	Content of antioxidant (%(*W*))	Relative molecular mass (M_n)
1	1	1	1	1	26,349
2	1	2	2	2	7922
3	1	3	3	3	8856
4	2	1	2	3	20,815
5	2	2	3	1	21,769
6	2	3	1	2	7498
7	3	1	3	2	6597
8	3	2	1	3	6446
9	3	3	2	1	7403
M_{1j}	14,376	17,920	13,431	18,507	
M_{2j}	16,694	12,046	12,047	7339	
M_{3j}	6815	7919	12,407	12,039	
R_j	9879	10,001	1384	11,168	

3.2.4 Orientation and crystallinity of the PLA fiber

3.2.4.1 In the electrostatic field

The orientation and crystallinity of the fiber have a significant effect on its mechanical properties. In 2016, Li et al. [29] studied the effects of speed and temperature of hot airflow on the macromolecular orientation and crystallinity of PLLA fibers, and the results indicated that the PLLA fibers with higher degrees of macromolecular orientation and crystallinity could be prepared under the condition of hotter airflow having faster speed. The experimental steps are as follows.

3.2.4.1.1 Materials and methods

The PLLA used in this study (with a molecular weight of 100,000 $g \cdot mol^{-1}$) was provided by the Zhejiang Haizheng Biomaterials Co., China. The PLLA had a glass transition temperature (T_g) of 61°C, a melting temperature (T_m) of 171°C, and a melt flow rate of 3–5 $g \cdot (10\ min)^{-1}$ at (190°C/2.16 kg).

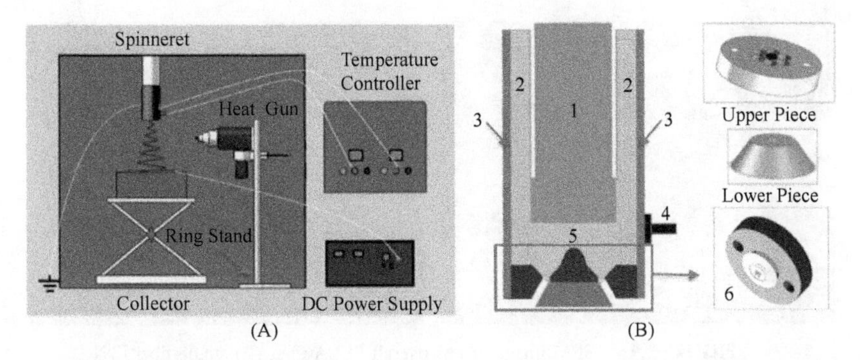

FIGURE 3.15 (A) A schematic representation showing the melt electrospinning apparatus with a hot airflow system, and (B) an illustration depicting the details of the spinneret (1. plunger; 2. barrel; 3. heating coil; 4. thermal couple; 5. polymer melt; and 6. frustum nozzle with upper and lower pieces)[29] .

The experimental setup for making PLLA fibers had two parts, including a hot airflow system and a melt electrospinning apparatus [30,41]. As shown in Fig. 3.15A, the hot airflow system consisted of a heat gun and a ring stand; this heat gun could produce two (i.e., fast and slow) speeds of hot airflow at two temperatures (i.e., 80 and 120°C). The melt electrospinning apparatus consisted of seven components including a plunger, a barrel, a heating coil, a thermocouple, a frustum nozzle, a high-voltage direct-current (DC) power supply, and a collector, while the frustum nozzle had two pieces as shown in Fig. 3.15B. During melt electrospinning, the molten PLLA flowed through six orifices in the upper piece and then spread onto the conical surface of the lower piece before being electrospun into filaments/fibers. The melt electrospinning was carried out at 200°C, the applied voltage was set at 40 kV, and the distance between the spinneret and collector was set at 18 cm.

The thermal characteristics of PLLA fibers were investigated using a DSC (Mettler-Toledo DSC 800), and a sample was heated from 20 to 200°C at the heating rate of 10°C · min^{-1} in nitrogen atmosphere. Based on the DSC results, the crystallinity was calculated from the following equation: $\chi_c = \frac{\Delta H_m - \Delta H_c}{\Delta H_0} \times 100\%$, where ΔH_m and ΔH_c are the enthalpies of melting and cold crystallization, respectively; while ΔH_0 is the heat of fusion of 100% crystalline PLLA (93 J · g^{-1}).

XRD was employed to examine the crystalline structures of PLLA fibers. XRD studies were carried out using a Rigaku D/Max 2500 VB2þ/PC X-ray diffractometer, and the X-ray tube was operated at 40 kV and 50 mA using the Cu-Kα radiation ($\lambda = 0.154$ nm). The XRD profiles were recorded in the 2θ range of 5°—35° at a scanning speed of 3° · min^{-1}.

3.2.4.1.2 Results and discussion

In this study, five types of PLLA fibers were prepared via melt electrospinning at 200°C with or without hot airflow, and the hot airflow could be set at 80 or

FIGURE 3.16 SEM image of (A) overall PLLA fiber (B) single fiber [29].

120°C with a slow or fast speed. In general, all melt electrospun PLLA fibers were morphologically similar and without significant discrepancies. Nevertheless, the PLLA fibers electrospun under the conditions of 120°C hot airflow with a fast speed appeared to be slightly thinner. Additionally, no beads and/or beaded fibers could be identified microscopically [42]. The SEM image in Fig. 3.16A shows the representative morphology of PLLA fibers prepared in this study. The fibers had diameters of several microns, and such diameters are common for melt electrospun polymer fibers; note that for the fibers electrospun from polymer solutions, the diameters are typically thinner than 1 mm. This is primarily because the viscoelastic force associated with a polymer melt is usually, in the order of magnitude, larger than that associated with a polymer solution. While the solidification rate of jets/filaments during melts electrospinning may be considerably faster than that during solution electrospinning, the fast solidification rate would also lead to the formation of polymer fibers with low crystallinity (i.e., the quenching effect) [43], particularly for PLLA that has a relatively slow crystallization speed. It is interesting to note that some fine PLLA fibers with diameters of 500 nm (Fig. 3.16B) were frequently identified, whereas the formation mechanism of these fine PLLA fibers is still under investigation.

Since the macromolecular orientation and crystallinity of melt electrospun PLLA fibers are typically low [43], their mechanical properties (particularly strength) are expected to be low as well. Such a situation may hinder/limit applications. Hence, it is important to improve the macromolecular orientation and crystallinity of melt electrospun PLLA fibers. Our hypothesis was that the introduction of hot airflow during melt electrospinning would increase the environmental temperature; thus the time period for the electric force to stretch the jets/filaments could be prolonged, which might lead to an improvement of the macromolecular orientation and crystallinity of PLLA fibers. Furthermore, the hot airflow had fast speed and it would be able to generate additional stretching force (similar to that in melt blowing), making the degrees of macromolecular orientation and crystallinity of the resulting PLLA fibers even higher. To verify

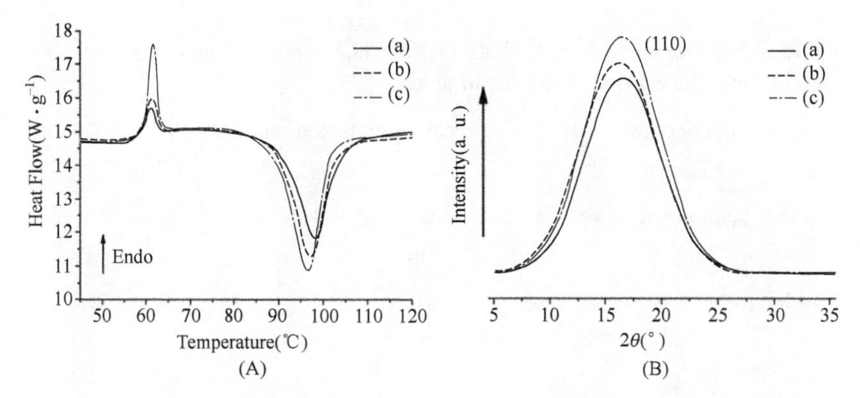

FIGURE 3.17 DSC (A) and XRD (B) curves acquired from the melt electrospun PLLA fibers (a) without hot airflow, (b) with 120℃ hot airflow having slow speed, and (c) with 120℃ hot airflow having fast speed[29].

the hypothesis, DSC and XRD analyses were performed; and the results are shown in Fig. 3.17; note that the melt electrospun PLLA fibers with an 80°C hot airflow having slow and fast speeds exhibited similar trend, while those curves are not included in Fig. 3.17 for clarity of presentation. As shown in Fig. 3.17A, the DSC curves acquired from three types of PLLA fibers had endothermic peaks (in 60–65°C) and exothermic peaks (in 85–105°C). Since the T_g of PLLA is ~ 61°C, the endothermic peaks are attributed to the relaxation of oriented (while not crystallized) PLLA macromolecules in the amorphous domains of fibers. It is evident that the melt electrospun PLLA fibers without hot airflow (curve (a)) had the smallest endothermic peak, indicating that their degree of macromolecular orientation was the lowest. The introduction of a 120°C hot airflow with a slow speed (curve (b)) increased the environmental temperature and prolonged the time period for stretching the electrospinning filaments/fibers. As a result, the degree of macromolecular orientation in those PLLA fibers was higher, thus the endothermic peak was larger. When the 120°C hot airflow with a fast speed was utilized (curve (c)), the additional stretching of filaments/fibers was achieved; in consequence, the degree of macromolecular orientation in the resulting PLLA fibers was higher, and the endothermic peak was larger. It was our speculation that, after the relaxation, the previously oriented PLLA macromolecules in the fibers would still possess a certain degree of loose alignment, which would facilitate subsequent crystallization (known as cold crystallization) in the temperature range of 85—105°C; and this speculation could be partially supported by the exothermic peaks in Fig. 3.17A. The different degrees of macromolecular orientation in three types of PLLA fibers could be further verified by XRD results; as shown in Fig. 3.17B, three XRD patterns had peaks centered at the 2θ value of ~ 17° that was attributed to the (110) crystallographic plane of PLLA [7]. Unlike the endothermic peaks in Fig. 3.17A, the XRD peaks in Fig. 3.17B originated from crystalline domains in the fibers. Evidently, the PLLA

TABLE 3.5 Degrees of crystallinity (χ_c) of PLLA fibers versus processing conditions during melt electrospinning [29].

Processing condition	Temperature of airflow (°C)	χ_c (%)
Without hot airflow	–	8.3
With hot airflow having slow speed	80	14
	120	18.6
With hot airflow having fast speed	80	20.5
	120	27.2

fibers melt electrospun with a 120°C hot airflow having a fast speed possessed the highest crystallinity, while the melt electrospun PLLA fibers without hot airflow possessed the lowest crystallinity.

To quantitatively study the effects of hot airflow on crystallinity of melt electrospun PLLA fibers, the crystallinity values were extracted from DSC curves using the equation in the experimental section, and the results are summarized in Table 3.5. As compared to the value of melt electrospun PLLA fibers without hot airflow, the crystallinity values of melt electrospun PLLA fibers with hot airflow were substantially higher. As explained before, the high environmental temperature would slow down the solidification of filaments/fibers during melt electrospinning, thus the stretching time period of electric force could be prolonged. With the increase of a hot airflow temperature from 80 to 120°C, the crystallinity values of PLLA fibers were also increased (with an effect similar to annealing) [43]. Furthermore, the hot airflow could apply additional stretching to the filaments/fibers. In this study, the hot airflow was perpendicular to the spinning direction due to the bending instability; the filaments/fibers would be in parallel to the hot airflow direction. Therefore, the hot airflow with a fast speed could effectively stretch the spinning filaments/fibers, leading to a high degree of crystallinity. For example, the crystallinity value (χ_c = 27.2%) of melt electrospun PLLA fibers with a 120°C hot airflow has a fast speed, which improved over three times compared to the crystallinity value (χ_c = 8.3%) of melt electrospun PLLA fibers without hot airflow.

3.2.4.1.3 Session conclusion

The introduction of hot airflow during melt electrospinning increased the environmental temperature, thus prolonging the time period for the electric force to stretch the spinning filaments/fibers; furthermore, the hot airflow with a fast speed could also result in additional stretching force. Therefore, the melt electrospun PLLA fibers under the conditions of a hotter airflow with a faster speed possessed higher degrees of macromolecular orientation and crystallinity. It is envisioned that the prepared melt electrospun PLLA fibers with relatively high macromolecular orientation and crystallinity (thus relatively high mechanical properties) would be particularly useful in future biomedical studies.

3.2.4.2 Effects of pulsed electric field

The experimental setup and procedure are consistent with those in Section 3.3.2.1, and the fibers are prepared for crystallization and orientation studies [28].

3.2.4.2.1 Crystallization behavior

In order to verify the above results, DSC, XRD, and polarized Raman measurements were carried out to further investigate the influence of pulsed electric fields on the internal structure of PLLA melt electrospun fibers, as shown in the following figures. The samples are: (a) as-received PLLA granules, (b) electrospun fibers prepared with melt electrospinning (control specimens), and electrospun fibers prepared with pulsed electrospinning with frequency and duty cycle, at (c) 1 kHz, 9.3%, (d) 1 kHz, 49.8%, (e) 4.03 kHz, 9.3%, and (f) 4.03 kHz, 49.8%. This nomenclature was used to identify these samples in the full text unless otherwise specified. The results of the DSC scans are compared in Fig. 3.18. As can be seen, all fiber samples had an enthalpy relaxation peak near T_g and a cold crystallization peak around 100°C,

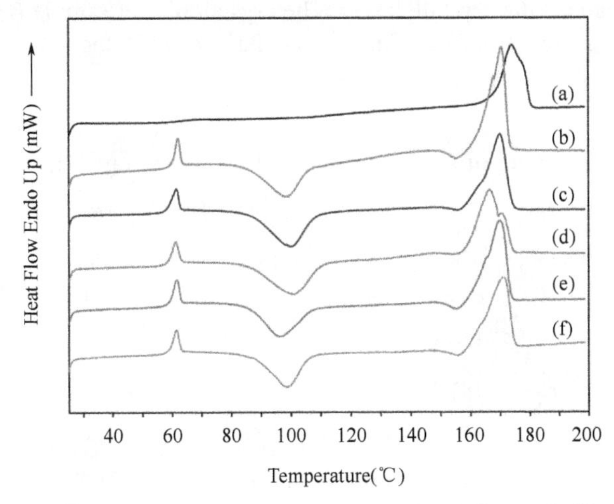

FIGURE 3.18 DSC scans of (a) PLLA granules and electrospun fibers with different conditions, (b) normal electrospinning, and (c) 1 kHz, 9.3%, (d) 1 kHz, 49.8%, (e) 4.03 kHz, 9.3%, and (f) 4.03 kHz, 49.8%[28].

suggesting that there exist amorphous sections and orientation structures in the melt electrospun fibers [44]. The T_g of fibers produced in the pulsed electric field were slightly lower than those of fibers produced using normal melt electrospinning, indicating that the chain movement was enhanced by the pulsed electric field in a subtle way. A small exothermic peak at 155°C occured in the fiber samples, which originates from the formation of imperfect α' crystal during the electrospinning process [45,46]. The melting peak of all samples shows an unobvious peak splitting, except sample (d). Melting peaks of DSC curve have been observed in many semicrystalline polymers, which were generally explained from two aspects; the presence of two crystal forms, and recrystallization effects during heating [47]. Zhou et al. [43] ascribed this phenomenon of PLLA fibers to different α and β crystal structures, while Song et al. [48] ascribed it to the melting of partial α'crystals and perfect α crystals. We will discuss this issue later with XRD analysis.

Table 3.6 summarized the numerical values of the glass transition temperature (T_g), cold crystallization temperature (T_c), melting temperature (T_m), cold crystallization enthalpy (ΔH_c), melting enthalpy (ΔH_m), and the degree of crystallinity X (%) obtained from the DSC curves. The crystallinity was calculated using Eq.(3.1) [49]:

$$X = \frac{\Delta H_m - \Delta H_c}{\Delta H_f} \times 100\% \tag{3.1}$$

ΔH_f is the heat of fusion for 100% crystalline PLLA, 93.7 $g \cdot J^{-1}$[48,50]. As shown in Table 3.6, the crystallinity of fibers produced using a pulsed electric field is still low (similar to normal electrospinning), but the results show that a high frequency (constant duty cycle) and a high duty cycle (constant frequency) enhanced the crystallization. The crystallinity of sample (f) is almost twice that of sample (b). According to Ramakrishna [51], the decreased T_c can

TABLE 3.6 Summary of DSC curves for different samples [28].

Sample	T_g (℃)	T_c (℃)	T_m (℃)	ΔH_c ($J \cdot g^{-1}$)	ΔH_m ($J \cdot g^{-1}$)	X (%)
a	58.4		174.0		60.4	64.5
b	62.1	98.6	170.5	53.0	58.5	5.9
c	61.5	100.1	170.0	42.4	47.8	5.8
d	61.3	101.2	170.9	39.6	47.2	8.1
e	61.7	96.4	169.8	46.3	53.6	7.8
f	61.4	98.9	171.0	40.7	51.4	11.4

a, PLLA granules; b, Normal melt electrospinning; c, 1 kHz, 9.3%; d, 1 kHz, 49.8%; e, 4.03 kHz, 9.3%; f, 4.03 kHz, 49.8%.

be attributed to a decrease in the conformational entropy of the molecular chains, due to the preferential orientation of the polymer chains caused by the electrospinning process. Therefore, we conclude that a duty cycle of 10% and frequency of 4 kHz can improve molecular orientation and then improve the crystallization rate of the fibers [52]. These results were in good agreement with SEM observations discussed previously (Fig. 3.19A and B). The molecular orientation is due to the fact that when a strong electric tensile force is applied in the form of a series of pulses on the polar functional groups of

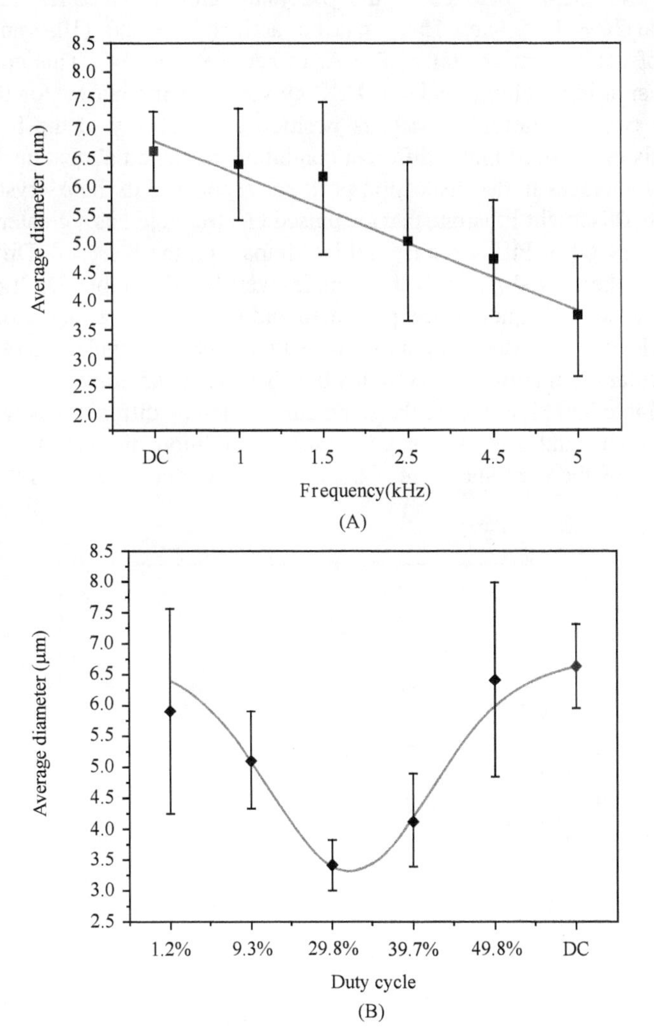

FIGURE 3.19 Relationship between the fiber diameter and (A) the frequency (duty cycle = 49.8%) and (B) the duty cycle (frequency = 1 kHz)[28].

PLLA molecular chains, the molecular chains are extended along the fiber axis. In fact, higher molecular orientation will increase the crystallization rate. However, a higher crystallization rate does not necessarily result in higher crystallinity [53].

3.2.4.2.2 Crystalline structure

To determine the crystal structures of the electrospun fibers, XRD studies were conducted, and the diffraction patterns are shown in Fig. 3.20 (beginning from (b), because (a) is connected to raw materials). It was observed that the weaker peak at $2\theta = 31.0^\circ$ belongs to the β-crystal, and the broader dispersion peak near $2\theta = 12.5^\circ$ and 15.0° can be assigned to the (103) and (010) planes of the α-form crystal of PLLA, respectively [48,54]. This confirmed that the splitting melting peaks in DSC curves were the reason for the existence of two polymorphs. The peak position and intensity of the β-form of PLLA fibers prepared under different conditions remained almost unchanged. However, changes in the dispersion peak corresponding to the α-crystal were apparent. This might illustrate that the pulsed electric field has a greater impact on the α-crystal of PLLA fibers, and less impact on the β-crystal. Diffraction peaks and the crystal plane of the samples were listed in Table 3.6. It showed that, at the same frequency, the peak area and intensity were increased as the duty cycle was increased, which means that the high duty cycle was beneficial to molecular orientation and crystallization. This is in agreement with the DSC data in Table 3.6. However, at the same duty cycle, the difference between the peak intensity and area was not obvious. In addition, the (103) and (010) reflections of the α-crystal were obtained when the duty cycle was 9.3% and

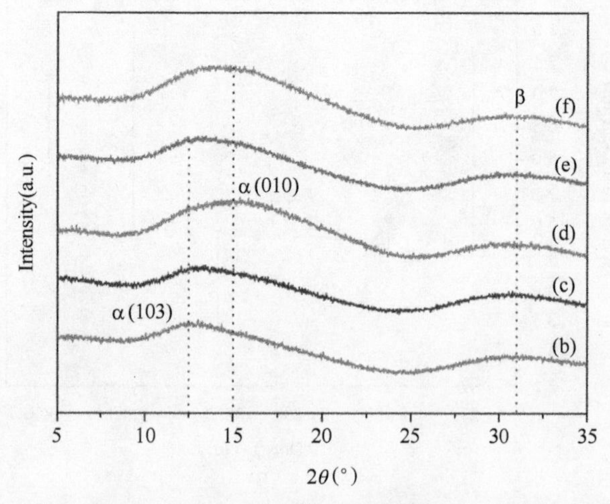

FIGURE 3.20 XRD patterns of electrospun fibers with different conditions: (b) normal electrospinning; (c) 1 kHz, 9.3%; (d) 1 kHz, 49.8%; (e) 4.03 kHz, 9.3%; and (f) 4.03 kHz, 49.8%[28].

49.8% at the same frequency, respectively. This was different from normal melt electrospun fibers. We noted that, when the frequency increased, the (103) diffraction peak was slightly shifted to a higher degree, while the (010) plane shifted to a lower degree, suggesting that high frequency results in smaller lattice spacing for (103) plane and larger lattice spacing for (010) plane according to the Bragg equation [55]. It should be pointed out that we cannot exclude the possibility that the imperfect α'-phase may also be presented in the fibers since the structure of α'-crystal is very similar to that of the α-crystal [50,56].

3.2.4.2.3 Molecular orientation

On the basis of the preceding SEM, XRD, and DSC findings, we believed that pulsed electric fields have a special effect on the molecular chains of PLLA electrospun fibers. Herein, the molecular chain orientation was characterized by polarized Raman spectra. The polarized Raman spectra from three different polarization geometries XX, YY, and YX according to Porto's notation [57] for the fiber samples (Fig. 3.6) are in the range of 800–1900 cm^{-1} range. It can be seen that three regions of the polarized Raman spectra were very sensitive to the conformational changes that took place during the electrospinning process: the skeletal stretching and CH_3 rocking band region of 800–950 cm^{-1}, the CH_3, CH bending, and C–O–C stretching band region of 1000–1500 cm^{-1}, and the C=O stretching band region of 1700–1900 cm^{-1}. Ribeiro et al. [58] reported that the 908 cm^{-1} absorption band is characteristic of the β-crystal and the 921 cm^{-1} absorption band is characteristic of the α-crystal. However, both of these characteristic bands are almost nonexistent in Fig. 3.21B–F. This may be due to the relatively low crystallinity of the melt electrospun fiber samples. Completely different intensities were obtained for the YY and XX spectra for all samples, indicating that the electrospun fibers (both from constant and pulsed electric field electrospinning) exhibit a high degree of anisotropy. To quantitatively assess the molecular orientation of crystalline and amorphous regions along the fiber axis, the depolarization ratio ρ was calculated using Eq. (3.2): [59].

$$\rho = \frac{I_{YX}}{I_{YY}} \tag{3.2}$$

I_{YX} and I_{YY} represent the intensity of Raman bands of the perpendicular and parallel polarization directions, respectively. The ρ value for a symmetric vibration mode 0.75 is $0 \leqslant \rho < 0.75$, whereas for an antisymmetric vibration mode ρ is equal to 0.75 [60]. In other words, when the depolarization ratio ρ of the sample is close to zero (the sample is anisotropic), the molecular orientation along the fiber axis is improved.

Four Raman bands were selected (Table 3.7) [57,58,61] to obtain the depolarization ratio, and the results were presented in Fig. 3.21F. It can be seen that the corresponding orientation parameters display similar trends across the

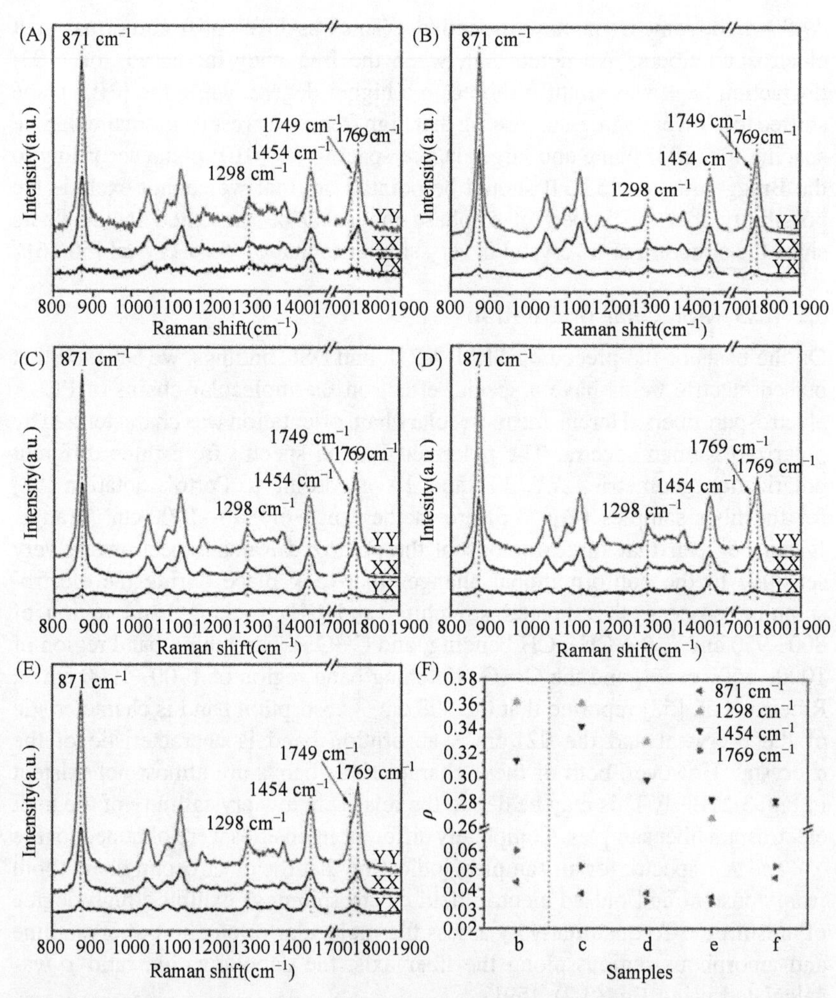

FIGURE 3.21 Polarized Raman spectra of the electrospun fibers produced at (A) normal electrospinning, (B) 1 kHz, 9.3%, (C) 1 kHz, 49.8%, (D) 4.03 kHz, 9.3%, and (E) 4.03 kHz, 49.8% for three different polarization geometries. (F) Depolarization ratio ρ of four Raman bands, with the letters of the horizontal axis identifying the corresponding samples [28].

TABLE 3.7 Vibrational and phase assignments of the Raman shift of PLLA[28].

Raman shift (cm^{-1})	Intensity	Assignment	Phase
871	Strong	υ(C–COO)	α
1298	Weak	δ(C–H)	Amorphous
1454	Strong	$\delta_{as}(CH_3)$	Amorphous
1769	Strong	υ(C=O)	α

different Raman bands. This reveals that both the normal and the pulsed electric fields make electrospun fibers exhibit a higher degree of molecular orientation. At the same time, the ρ value of the 871 and 1769 cm^{-1} Raman bands is lower than from the 1298 and 1454 cm^{-1} bands, suggesting that the molecular orientation of the electrospun fibers should be better in the crystalline region than in the amorphous region. However, the highly ordered molecules does not contribute much to the crystallinity (see Table 3.6 and Fig. 3.20). Because of the strong electric force acting on the jet and its extremely rapid solidification (which can be compared to a fast quenching process), the jet typically has no time to form stable crystals. This behavior could also explain the DSC and XRD curves. More importantly, the results show that, in a certain range, the pulsed electric field is more beneficial to the molecular orientation of the electrospun fibers, compared to the constant electric field. At low frequencies, better molecular orientation results from higher duty cycles (see samples c and d), while at higher frequencies, better molecular orientation results from low duty cycles (see samples c and e). However, at a frequency of 4 kHz (which is relatively high), the molecular orientation deteriorates as the duty cycle increases (see samples e and f). These results suggest that the synergistic effects of high frequency and high duty cycles may lead to serious local stress concentration on the jet; it disperses the polymer network faster and more loosely, which is more conducive to a fully stretched configuration and is helpful for chain disentanglement. On the basis of the above discussion, the stretching and orientation changes of PLLA molecular chains during normal electrospinning and pulsed electric field electrospinning were shown schematically in Fig. 3.22. However, due to network separation, the tension in the chains close to the breaking point is

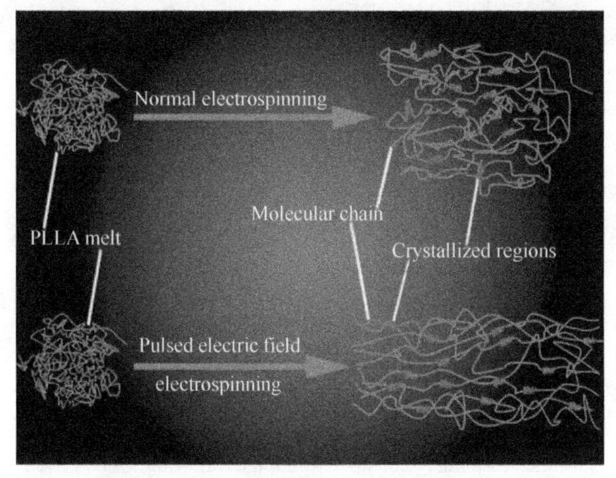

FIGURE 3.22 Schematic of stretching and orientation changes of PLLA molecular chains during normal electrospinning and pulsed electric field electrospinning [28].

relieved, and their relaxation leads to regaining some of the disentanglement when the pulse voltage reaches zero [62]. Accordingly, at high frequencies and high duty cycles, the degree of molecular orientation is reduced, and the fiber diameter is decreased.

A phenomenon we have to mention is that unexpected multiple necking structures were detected at frequencies of 8 and 10 kHz (Fig. 3.23A–D), while the fibers are smooth and have uniform morphology under the same conditions of a normal electrospinning process (Fig. 3.23E). These unexpected features have been reported in solution electrospinning under a constant electric field [63–65]. However, we obtained similar structures when using pulsed electric

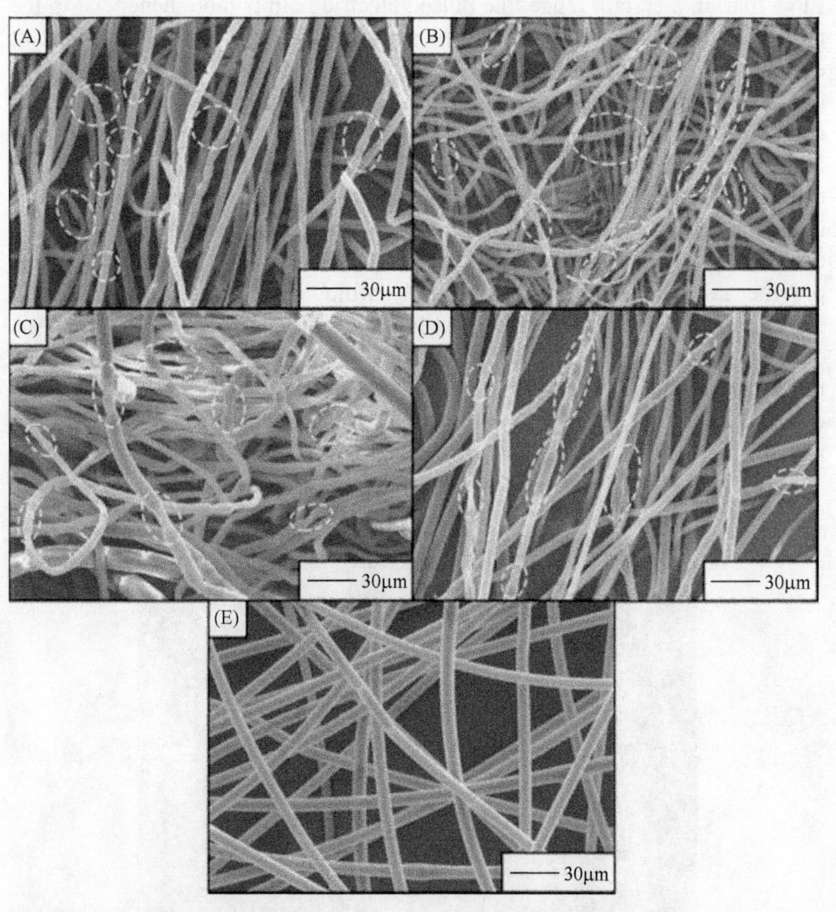

FIGURE 3.23 SEM images. Some obtained fibers have special multiple necking structures of pulsed electric field electrospinning, marked by yellow oval frames in the pictures. The frequency and duty cycle of the electrospinning conditions were the following: (A) 8 kHz, 20%; (B) 8 kHz, 30%; (C) 10 kHz, 10%; and (D) 10 kHz, 40%. Image (E) shows the smooth and uniform fiber morphology of normal electrospinning [28].

fields with a stationary flat collector in melt electrospinning. Similar to these observations, the necks in a single fiber in the present study are not evenly spaced along the fiber axis. The fiber collection and sample preparation under all experimental conditions are the same and do not make any changes artificially, excluding other possible reasons for multiple neck deformation. Obviously, the multiple necking structures in this study can be directly related to the frequency and duty cycle of the pulsed electric field. At first glance, the nature of this interaction appears to be a complicated one, which requires an explanation in a future study. Generally, it is believed to be related to the following aspects: (1) High-frequency pulsed electric fields provide a huge energy accumulation, causing large differences between the on and off voltage periods over one cycle. (2) The electrically driven jet or melt flow in the electrospinning process will experience nonaxisymmetric instability and two axisymmetric instabilities, leading to the formation of multipoint local stress concentration on the jet and further propagation of this fluctuation. (3) The high-frequency pulses contribute more on the dispersion of the polymer network, which gives rise to a complex relationship between the electric force, disentanglement, and jet solidification. There may be a critical entanglement concentration analogous to the entanglement concentration Ce in solution electrospinning proposed by Zussman [62] and McKeen [66] for this system, which requires further study in order to be determined.

3.2.4.2.4 Session conclusion

Increasing the frequency and duty cycle can improve the molecular orientation and crystallinity of the fibers. However, special multiple necking structures are observed at frequencies of 8 and 10 kHz. This may be related to the entanglement concentration, jet stretching, and solidification. Generally, pulsed electric field electrospinning is expected to break through the fiber diameter in the bottleneck currently present in the field of melt electrospinning. It could also improve the performance of such fibers and lead to the development of special fiber structures. Pulsed electric field electrospinning has great potential and deserves further attention and research.

3.3 Phenolic resin

Phenolic fibers were invented by Economy in 1972 and used for aerospace insulation materials [67]. Owning to their extraordinary flame, heat and corrosion resistance, high carbonation rate, high carbon yield, as well as cheap and readily available raw materials, phenolic fibers have been widely used in many fields, such as bunker clothing, fire-proof plates, antifriction components, corrosion-resistant fabric, sound insulation, composites, and as precursors for carbon fibers and activated carbon [68–71]. Phenolic fibers are usually prepared by solution spinning or melt spinning. However, the obtained fibers always have a large diameter and the preparation process is very complicated [72,73]. With the development of modern technology, the need for thinner flame-retardant fibers is ever-increasing for nano- and microelectronic devices. Phenolic fibers are a good candidate to meet this requirement.

Electrostatic spinning technology has not only been more mature in recent decades, but many researchers have combined electrospinning technology in various studies [74–76]. In recent years, phenolic fibers are mainly prepared by solution electrospinning [77–79]. These studies have greatly contributed to the development of phenolic fibers. However, the preparation of phenolic fibers by solution electrospinning still has some drawbacks. First, the residual solvent in the fiber can cause defects in the fiber and affect its physical properties. Most of the solvents are generally dimethyl formamide (DMF) and methanol and other toxic substances, and in the spinning process, a large number of volatile substances, not only pollute the environment, but are also harmful to human health. Second, the phenolic resin is a low-molecular-weight polymer, and it is necessary to add polyvinylpyrrolidone (PVP), polyvinyl alcohol (PVA), and other carriers to adjust the viscosity of the solution. There is also the need to add a certain amount of Na_2CO_3, pyridine, ethanol, surfactant, PVP, PVA, and other [69,70,77,78,80,81] additives to improve the viscosity of the spinning solution and conductivity. It can be seen that the process is very complex. Moreover, there are very limited types of phenolic fiber additives currently used for solution electrospinning. Finally, the fiber yields are very low in solution electrospinning. However, as far as we know, only Gee's patent describes the preparation of phenolic fibers by melt electrospinning. The patent relates to the melt phenolic polymerization system, the molecular weight structure, and properties of the resulting fibers, which are discussed in detail. The electrospinning of the melt in the absence of a solvent drawn to the fibers from the heated polymer is considered to be a more environmentally friendly and simpler process than solution electrospinning, which avoids contamination and damage to toxic solvents. In addition to increased production, the fibers have fewer defects and better physical and mechanical properties.

According to the above analysis, based on orthogonal design experiments, our group [41] used the melt electrospinning to prepare pure phenolic fibers. Similarly, fiber properties such as crystallization and heat resistance were evaluated. The following describes the specific experimental steps.

3.3.1 Materials and equipment

The analytical-grade powder of para-tert-butylphenol (PTBP) formaldehyde resin was used without further purification as the raw material for melt electrospinning. The resin was purchased from Letai Chemical (China). The weight average molecular weight $(\overline{M_w}) = 2733\ g \cdot mol^{-1}$ and the number average molecular weight $(\overline{M_w}) = 1968\ g \cdot mol^{-1}$ of the received PTBP formaldehyde resin were measured by gel permeation chromatography (GPC; Waters 1525,

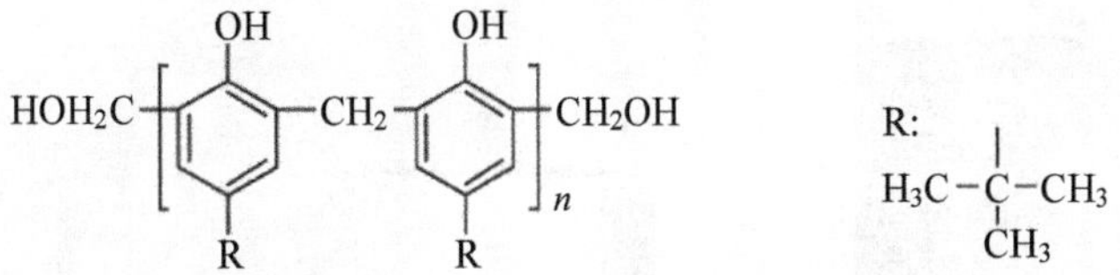

FIGURE 3.24 The formula of para-tert-butylphenol formaldehyde resin [41].

TABLE 3.8 Specification table of experimental materials and reagents.

Raw materials and reagents	Manufacturer	Remarks
PTBP	Tianjin Loctite chemical Co., Ltd	$M_w = 2733\ g \cdot mol^{-1}$
		$M_n = 1968\ g \cdot mol^{-1}$
Formaldehyde	Tianjin Daming chemical Reagent Factory	Analysis of pure
Hydrochloric acid	Beijing Modern Oriental Technology Development Co., Ltd.	Analysis of pure
Anhydrous ethanol	Beijing Modern Oriental Fine Chemicals Co., Ltd.	Analysis of pure
Distilled water	Laboratory homemade	Analysis of pure

USA). The molecular structure is shown in Fig. 3.24. The solvent was analyzed by tetrahydrofuran (THF) by GPC. The column temperature was 35°C, the injection volume was 50 mL, and the flow rate was 2.5 μL · min^{-1}.The melt electrospinning equipment used in this experiment is consistent with the article published in this group [30,35]. Other experimental materials and reagents are listed in Table 3.8. The spinning process is shown in Fig. 3.25.

The PTBP used in this experiment is very brittle, and has been partially broken into powder. Prior to the experiment, the raw materials were dried in a vacuum oven (DZF-6020, Shanghai Jie Experimental Instrument Co., Ltd.) at 60°C for 3 h, and then crushed into a homogeneous powder. In order to improve the brittleness of the phenolic fibers obtained by electrospinning of the melt, the phenolic fibers were placed in a mixture of formaldehyde and 12% hydrochloric acid, respectively, and then placed in a vacuum oven for crosslinking curing.

3.3.2 Orthogonal experimental arrangements

In this experiment, a $L_9(3)^4$ orthogonal table was used to arrange the experiment by orthogonal design assistant v3.1. The orthogonal table is named with a

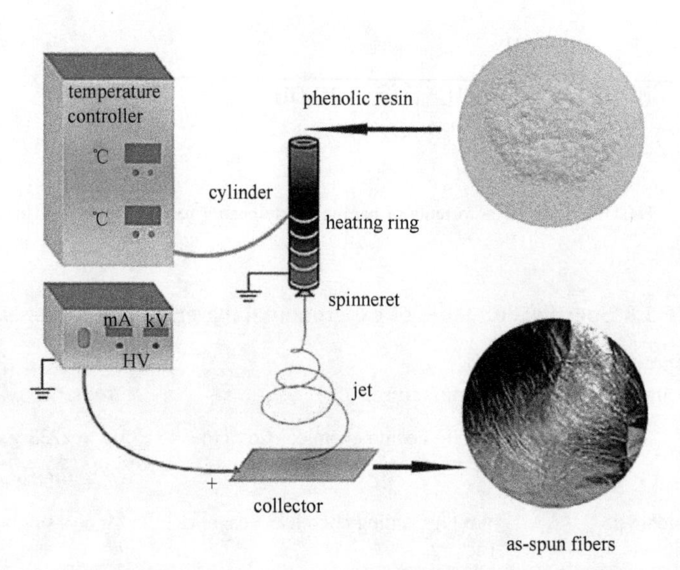

FIGURE 3.25 Schematic diagram of the melt electrospinning process [41].

specific mark $L_n(t)^c$. L represents the orthogonal table, t is the number of levels of each factor, the superscript c is the maximum number of factors allowed, and the subscript n represents the total number of experiments. In this experiment, three typical influencing factors were studied, spinning temperature, spinning distance, and spinning voltage, and labeled as A, B, and C [82].

It was found that when the temperature was lower than 140°C, the viscosity of the phenolic resin after melting was so high that the spinning efficiency was too low. When the temperature was higher than 160°C, the phenolic resin in the spinning process had undergone a serious thermal decomposition. Taking into account the high strength of the electric field required for melt electrospinning, in order to avoid the formation of high-precision spinning voltage safety hazards, there is a comprehensive selection of distance and voltage data, the final factor level of the experiment as shown in Table 3.9.

TABLE 3.9 Factors and levels for orthogonal experimental design [41].

	Factors		
Levels	A(°C)	B(cm)	C(kV)
1	140	8	30
2	150	9	35
3	160	10	40

(1) The molecular weight of PTBP was determined by GPC (Waters 1525), eluent, tetrahydrofuran at 35°C, volume 50 μL, and flow rate 2.5 μL · min^{-1}.
(2) Characterization of fibers: SEM (Hitachi SU510, Japan), each sample was subjected to surface metallization before performing a scanning surface test.
(3) Fiber diameter test: The diameter of the fiber on the sample SEM photograph was measured by software Image J. Twenty different fibers were selected for each sample and five different points were selected for each fiber. The data were analyzed by Origin to obtain the average diameter and diameter distribution values for orthogonal test for data analysis.
(4) Molecular structure characterization: Fourier transform infrared spectroscopy (FTIR, USA Thermo Fisher, Nicolet6700), resolution 0.1 cm^{-1}, temperature conditions for room temperature.
(5) Thermal performance analysis: thermogravimetric analyzer (TGA, Germany Netzsch Q500), temperature range 40–800°C, heating rate 10°C · min^{-1}, nitrogen environment (flow 50 mL · min^{-1}).
(6) Crystalline behavior: DSC (US Perkin Elmer Pyris 1), aluminum crucible, temperature range 25–130°C, heating rate 10°C · min^{-1}, nitrogen environment (flow 20 mL · min^{-1}).
(7) Crystallization analysis: XRD (Germany Bruker, D8 Focus), scanning range $2\theta = 4°-60°$, Cu Kα target, $\lambda = 1.5406$ Å (1Å=10^{-10}m), voltage 40 kV, current 40 mA. The resolution and scanning steps are 0.02° and 0.1 s·$step^{-1}$, respectively.

3.3.3 Optimal spinning conditions

The mean diameter (AD) and diameter distribution (SD) of the fiber were chosen as the evaluation criteria for the orthogonal test. K_1, K_2, and K_3 in Table 3.10, respectively, represent the average of the three levels of assessment factors corresponding to each factor. For example, the two-level AD value of factor B, that is, K_2, can be calculated as $K_2 = (21.36 + 3.42 + 7.04)/3 = 10.61$. By comparing the K value, the optimal level of each factor can be determined. R in Table 3.10 represents the difference between the maximum value of the evaluation index and the minimum value of K. For example, the R value of the average diameter of factor A can be calculated as $R = 15.74–7.79 = 7.95$. The difference R reflects the degree of benefit or disadvantage to D and SD. The greater the factor R, the greater the effect of the factor on the evaluation index, and the more important it is to understand [83]. Because the fiber diameter is more uniform and more conducive to its multifaceted performance, therefore, the K value in this experiment is as small as possible. The results of the analysis are shown in the lower part of Table 3.10.

TABLE 3.10 Result and analysis of orthogonal $L_9(3)^4$ experimental design [41].

	Factors			Results	
Experiment	*A* (°C)	*B* (cm)	*C* (kV)	Average diameter (μm)	SD of average diameter
1	1	1	1	15.43	13.02
2	1	2	2	21.36	2.03
3	1	3	3	10.44	1.58
4	2	1	2	6.36	0.92
5	2	2	3	3.42	2.85
6	2	3	1	18.64	8.94
7	3	1	3	4.44	0.76
8	3	2	1	7.04	4.06
9	3	3	2	11.88	2.41

Optimal conditions of orthogonal $L_9(3)^4$ experimental design	*A*	*B*	*C*	*A*	*B*	*C*
K_1	15.74	8.74	13.70	5.54	4.90	8.67
K_2	9.47	10.61	13.20	4.24	2.98	1.79
K_3	7.79	13.65	6.10	2.41	4.31	1.73
R	7.95	4.91	7.60	3.13	1.92	6.94
Order of importance		$A>C>B$			$C>A>B$	
Optimal level	A3	B1	C3	A3	B2	C3

SD means standard deviation;
$K_i^F = (\sum$ the value of evaluation indexes at the same level of each factor$)/3$;
$R^F = \max\{K_i^F\} - \min\{K_i^F\}$($F$ stands for factor A, B and C. i stands for levels 1, 2, and 3).

For a more intuitive representation of the results, Fig. 3.26 is plotted against the K values in Table 3.10. It can be seen from the figure that the optimum spinning conditions are A3 (160°C), B1 (8 cm), and C3 (40 kV). It is worth noting that, in this experiment, as the spinning distance increases, the fiber diameter also increases, which is contrary to the traditional single-factor electrospinning experiment. A possible explanation for this phenomenon is that orthogonal experiments are a comprehensive study of multivariate level research methods, unlike single-factor experiments in single-variable studies. In the orthogonal experiment, as the spinning distance becomes larger, the electric field strength may be reduced and the melt viscosity may increase, so that the resulting fiber diameter may become larger as the distance increases.

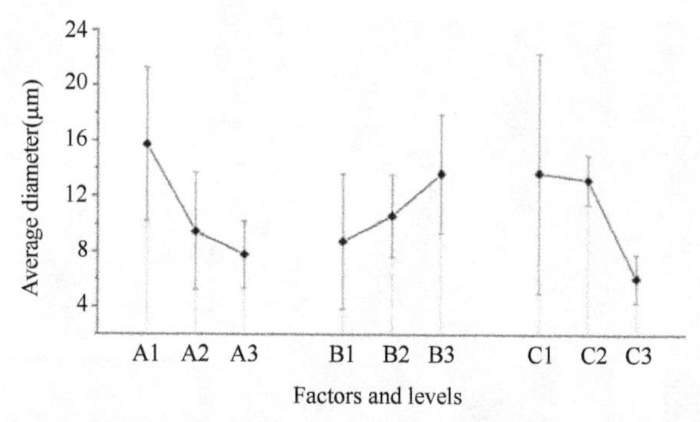

FIGURE 3.26 Relationship between average diameter with the three factors and levels[41].

Fig. 3.27 shows SEM images of the nine experiments in Table 3.10, and the finest fibers corresponding to each experiment are given. The right side of the histogram shows the fiber diameter distribution for each experiment. It can be seen from the melt electrospinning method that surface smooth phenolic fiber can be successfully obtained. The fiber diameter under the most favorable conditions is (4.44 ± 0.76) μm, although it is not the finest in the field of melt electrospinning, but this is much better than the long-haired melt-spinning phenolic fibers [72,84].

3.3.4 Fiber heat resistance and crystallinity

As can be seen in Fig. 3.28, the strength of the aliphatic characteristic peak C—H (2959 cm^2) of the vacuum-dried phenolic resin (curve (b)) is weaker than that of the raw material (curve (a)). The reason is that after the vacuum drying, the free phenol in the feedstock and other small molecules are removed. In the phenolic fiber (curve (c)), the strength of the aliphatic characteristic peaks C—H (1959 cm^{-1}) and the methylene bridge $-CH_2-$ (1488 cm^{-1}) were both weakened and the cured phenolic fibers (curve (d)) became strong. In addition, the aromatic bond CH tensile vibration (3049 cm^{-1}), 1,4- and 1,2,4-benzene ring (820 cm^{-1}) and 1,2- and 1,2,6-benzene ring (735 cm^{-1}) strength were weakened. Also, the phenolic hydroxyl-OH characteristic peaks moved to a higher wavenumber (3396 cm^{-1}). This phenomenon means that the melting temperature of the cured phenolic fibers is increased. A dehydration condensation reaction may occur between phenolic-OH (s) and methylene, resulting in a decrease in the strength of the phenolic hydroxyl group.

Fig. 3.29 shows the thermogravimetric analysis (TGA) and derivative thermogravimetry (DTG) curves for the sample. The corresponding initial decomposition temperature and 50% weight loss are T_{on} and $T_{0.5}$, respectively. The maximum decomposition rate temperatures T_{d1}, T_{d2} and the carbon content at 800°C obtained from the DTG curve are shown in Table 3.11. T_{d1} and T_{d2} of all samples do not change significantly, but the phenolic fiber T_{on} and $T_{0.5}$ and the amount of residual carbon are the lowest and curing the highest phenolic fiber. This indicates that the thermal decomposition rate of phenolic fibers is

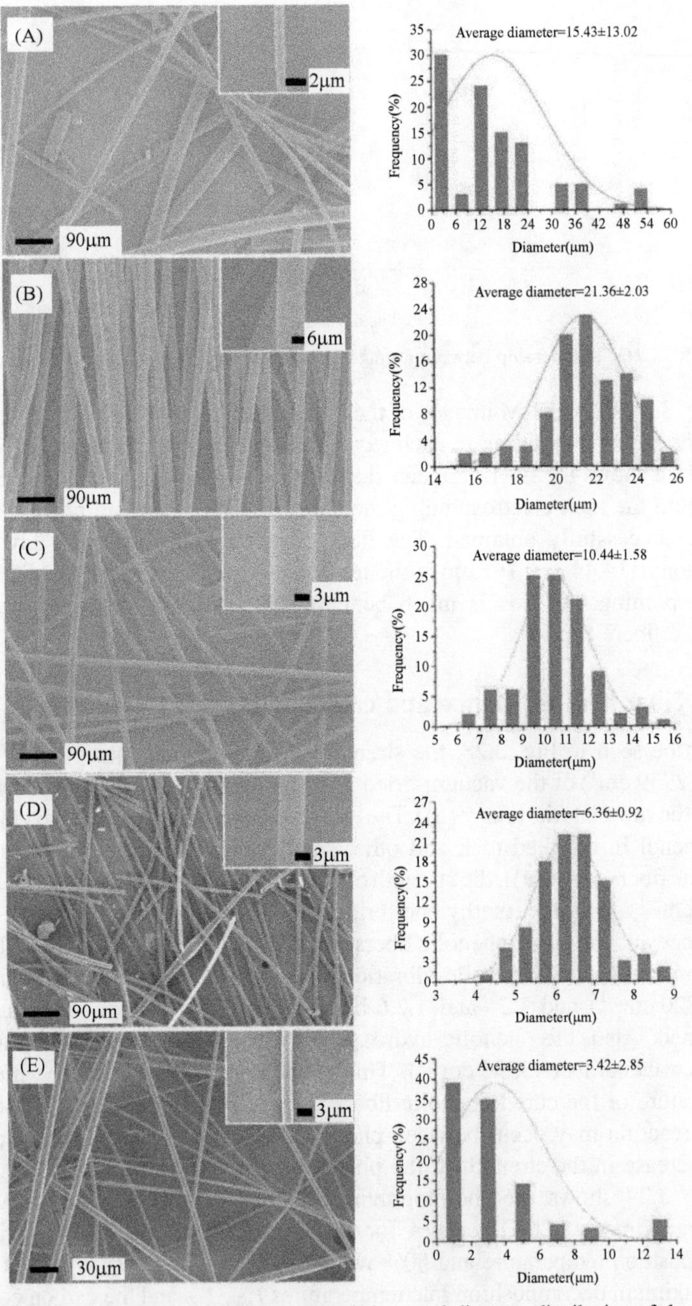

FIGURE 3.27 SEM images showing the morphology and diameter distribution of the as-spun phenolic fibers; (A)−(I) correspond to experiments 1−9, the histograms on the right correspond to the diameter distribution of each experiment[41].

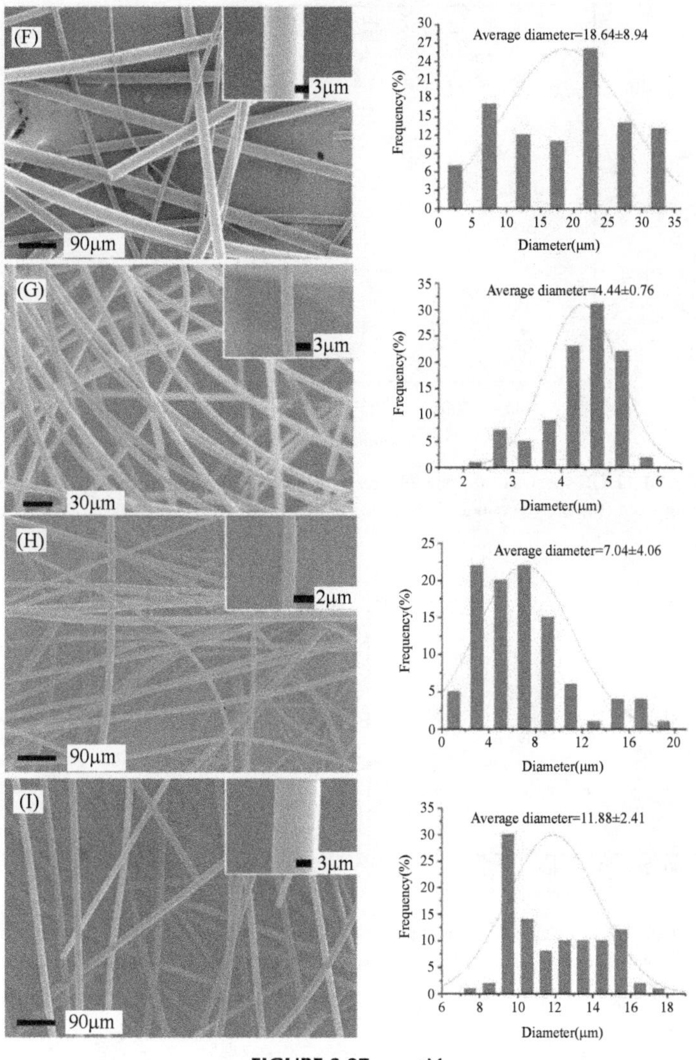

FIGURE 3.27 cont'd

higher than that of cured phenolic fibers. According to Bai's report [78], the difference in fiber diameter has a significant effect on the thermal stability of the fiber. With the decrease in fiber diameter, the heat loss during the thermal decomposition process is more pronounced, so the TGA of the phenolic fiber exhibits different thermal behavior due to the change in the size of the fiber. However, the melt electrospinning process may weaken some of the molecular bonds and these molecular bonds will break when heated again, resulting in different TGA characteristics. The above results show that the heat resistance of the phenolic fiber after curing has improved.

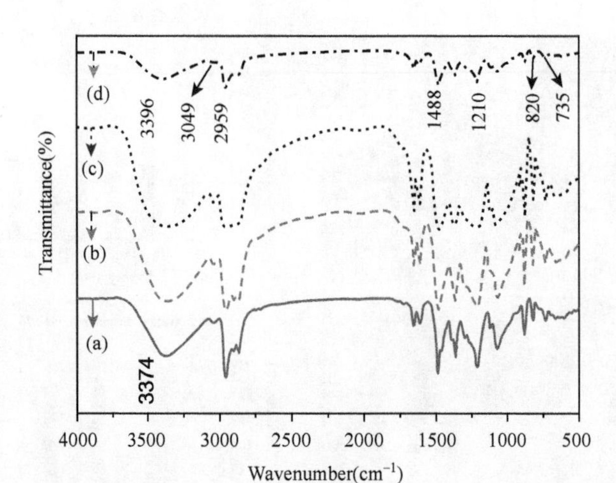

FIGURE 3.28 FTIR spectra of (a) phenolic resins, (b) phenolic resins dried in vacuum, (c) fibers electrospun from (b), and (d) cured fibers from (c)[41].

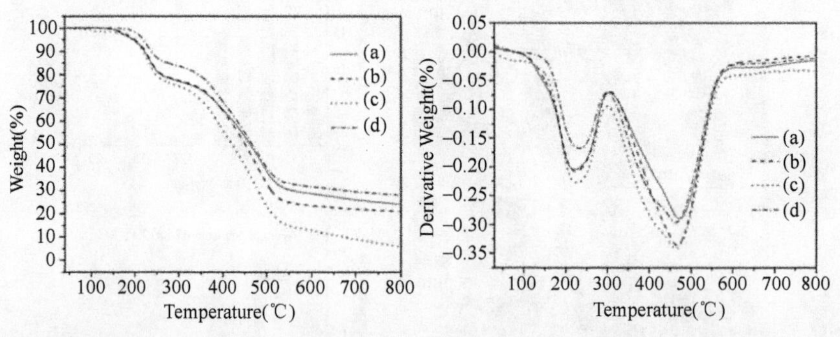

FIGURE 3.29 TGA analysis (left) and corresponding DTG curves (right) of (a) phenolic resins, (b) phenolic resins dried in vacuum, (c) fibers electrospun from (b), and (d) cured fibers of (c)[41].

TABLE 3.11 Thermal characteristics of phenolic resins and fibers[41].

Curve	T_{on} (°C)	$T_{0.5}$ (°C)	T_{d1} (°C)	T_{d2} (°C)	Residue at 800°C (%(w))
(a)	150	470	221	475	24.0
(b)	150	456	221	468	21.0
(c)	76	425	226	466	6.5
(d)	185	474	230	465	28.5

(a) phenolic resins; (b)phenolic resins dried in vacuum; (c)fibers electrospun from(b); (d)cured fibers of (c).

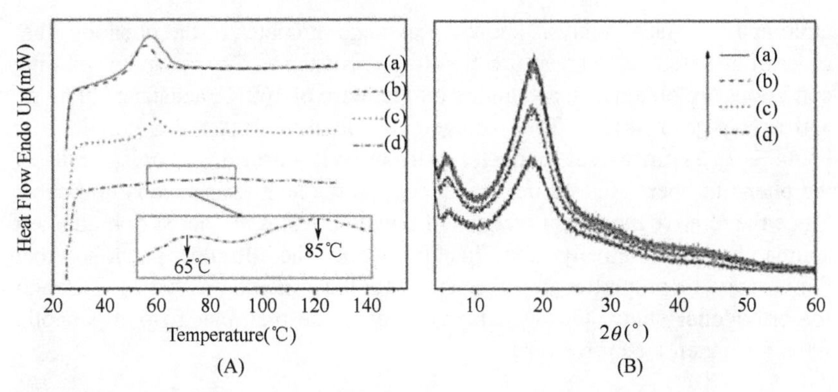

FIGURE 3.30 (A) DSC curves and (B) XRD curves of (a) phenolic resins, (b) phenolic resins dried in vacuum, (c) fibers electrospun from (b), and (d) cured fibers from (c) [41].

As shown in Fig. 3.30A, the melting temperature T_m of the phenol resin (b), after vacuum drying, is slightly lowered and the phenolic fiber (c) is increased. This may be due to the fact that the raw materials used in this experiment are unmodified p-tert-butyl phenolic resins containing some small molecules and monomers that may take on the role of "plasticizer" in the feedstock. However, these "plasticizers" are volatile in the melt electrospinning process, due to high-temperature heating, leaving long-chain molecules. Thus, the relative content of the long-chain molecules increases, resulting in an increase in the T_m of the phenolic fibers [85]. This peak is flat because the heat of the cured portion is exothermic and heats up. The melting peak at 85°C is due to the recrystallization of imperfect crystal structures [86]. This melting peak of the cured phenolic fiber (d) is much smaller than the other three samples ((a), (b), and (c)), indicating that the crystallinity of the phenolic fibers decreases after curing. And the heat resistance of the surface-cured phenolic fiber (d) is improved by increasing the melting peak temperature. The crystallinity based on the XRD diffraction curve (Fig. 3.30B) is obtained by software MDI jade 5.0 with an error of less than 9%. The final crystallinity data take the average of the three results. The crystallinities of samples a, b, c, and d is 15.86%, 15.50%, 13.4%, and 11.73%, respectively. The reduction of the relative strength of the phenolic fiber (c) on the XRD curve also shows that the melt electrospinning process destroys the regularity of the molecular chain and reduces the crystallinity. In addition, the decrease in the crystallization peak of the cured phenolic fiber (d) indicates that the curing process further reduces the crystallinity.

3.3.5 Session conclusion

In this experiment, the phenolic fiber is successfully prepared by melt electrospinning, and the electrospinning experiment is carried out by the orthogonal design method and was optimized. The orthogonal experimental table and the visual analysis show the average diameter of the phenolic fiber order of importance, temperature > voltage > distance. The optimum spinning conditions are obtained at a spinning temperature of 160°C, a distance of 8 cm, and a voltage of 40 kV. The average fiber diameter under this condition is (4.44 ± 0.76) μm, and the diameter distribution is uniform. The crystallinity of the phenolic fibers after curing is reduced but the heat resistance is improved. Since the relative molecular weights of phenolic resins are not very high, their spinnability is relatively low. In this work, the ultrafine phenolic fiber flame-retardant properties and other aspects of the in-depth are studied. Through the orthogonal study, the available quality of the melt electrospun phenolic fiber has taken a step forward.

3.4 Polypropylene (PP)

Since the Italian Montecatini company in 1957 achieved the commercialization of polypropylene production, it has become the plastic marketed with the most robust varieties [87]. PP has a melting temperature of 160–170°C, which is the highest among all thermoplastics materials and it has better heat resistance than other thermoplastics materials at low cost. Polypropylene is a semicrystalline thermoplastic material with a low cost, a low density, a high softening temperature, excellent mechanical properties, good chemical resistance, easy processing, and recycling. Due to its versatile nature, PP can be used in the manufacture of fiber, film, electric insulators, containers, medical equipment, mechanical parts, and an injection of household consumer goods, which are all important applications of polypropylene. Polypropylene has a high melting point, strength, and toughness. Additionally, when mixed with the proper additives, PP resins can show excellent processability. There are many ways to prepare the fiber, including electrospinning, melt spinning, phase separation, and other methods. As the fiber diameter decreases, the specific surface area of the material increases. Compared with the traditional fiber preparation method, electrospinning is the only way to achieve the continuous preparation of nanofibers. Ultrafine polypropylene fiber with a small density, high strength, chemical resistance, large surface area, and other unique excellent performances, has a very wide range of applications, such as construction, biomedicine, industrial filtration, military protection, and aviation engineering [88–93]. Since polypropylene is almost insoluble in organic solvents at room temperature, it is necessary to prepare polypropylene fibers by melt electrospinning [90,94]. At present, the study of melt electrospinning at home and abroad is still in the initial stage, most with the use of self-built devices, selecting a polymer for trial spinning. Research into the parameters (voltage, distance, temperature, etc.) of the electrospinning of the melt is not complete, and the optimal configuration between the process parameters needs to be further improved [95].

3.4.1 Equipment

In the laboratory group, the preparation of PP ultrafine fibers by melt electrospinning was studied in detail, including the process parameters, control of the spinning path, and modification of the polypropylene.

In 2010, our research laboratory [96] designed a self-made efficient spinning nozzle, as shown in Fig. 3.31, for the orthogonal experimental study of melt electrospinning PP fiber diameter factors. It provided a reference standard for future melt spinning experiments. Specific experimental steps are as follows.

Three grades of PP (PP6315, PP6312, and PP6310) were purchased from Shanghai Expert in the Developing of New Material Co. Their Melt mass-flow rate (MFR) values were 1500 g·$(10\ min)^{-1}$, 1200 g·$(10\ min)^{-1}$, and 1000 g·$(10\ min)^{-1}$, respectively. This experiment uses a patented device [96], as shown in Fig. 3.32. The collecting device is composed of a mesh cut to $10 \times 10\ cm^2$. The spinning conditions are maximum output voltage +60 kV, maximum output current of 2 mA, and the customized electric heating ring power is 300 W.

Initially the cylinder is heated to the desired temperature required for PP (i.e., 230°C) and its temperature is regulated by the temperature control system. When the temperature reaches a steady state PP particles are added to it. After 10 min, the PP particles are melted, and then high-power voltage is turned on and adjusts to the desired value to control the melt flow rate by pushing the piston. Under the action of the electric field force, the melt was ejected from the nozzle and the Taylor cone is formed. The electrospinning fiber morphology is observed and measured SEM (Hitachi S4700) and the fiber diameter is measured. Prior to observation, the fiber sample was plated with a platinum layer. In the case where the spinning conditions are the same, the average fiber diameter and the standard deviation are calculated.

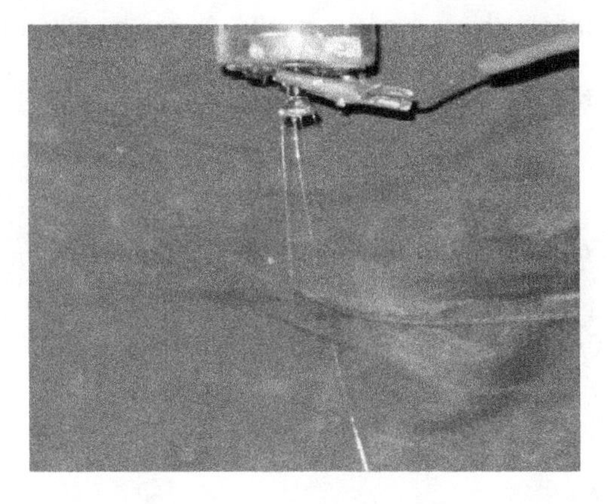

FIGURE 3.31 A high-efficiency spray head setting for electrospinning [96].

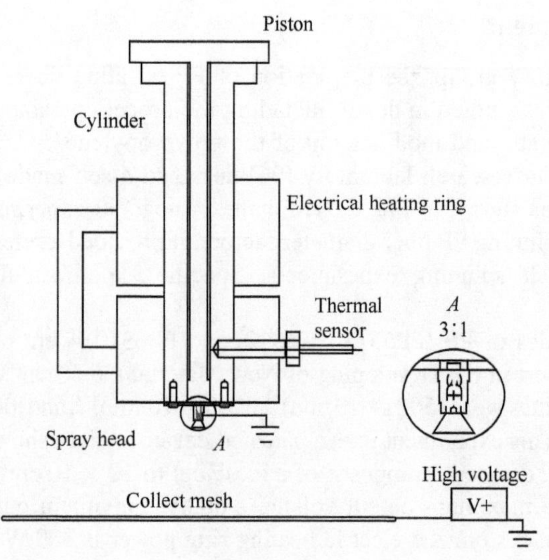

FIGURE 3.32 Schematic illustration of the experimental apparatus[96].

The measured results are analyzed by orthogonal experiments, as shown in Tables 3.12 and 3.13. Each of the four factors of temperature (A), voltage (B), distance (C), and MFR (D) can be changed at three levels. The effects of these four factors on fiber diameter and diameter distribution were tested. Table 3.12 shows the three levels of detail for each factor. In total, nine experiments were carried out according to the orthogonal array L_9 $(3)^4$. The influence of the four factors on the fiber diameter is shown in Table 3.13. M_{1j} represents the sum of the results of the j row when the level is 1. R_j represents the difference between the maximum and minimum M_{ij} values in the j row. The greater the R_j value, the greater the effect of factor j on the fiber diameter. At the same time, T corresponds to the sum of the results. As can be seen from Table 3.13, $49.41 > 46.16 > 44.65 > 41.21$, the influence of the four factors on the diameter can be in the following order: MFR > voltage > distance > temperature.

TABLE 3.12 Contents of orthogonal factors and levels[96].

Factor level	A Temperature (°C)	B Voltage (kV)	C Distance (cm)	D MFR (g·(10 min)$^{-1}$)
1	250	60	9.5	1500
2	230	50	13	1200
3	210	40	16.5	1000

TABLE 3.13 Effect of four factors on fiber average diameter and SEM figures[96].

Experiment	*A* Temperature (°C)	*B* Voltage (kV)	*C* Distance (cm)	*D* MFR $(g\cdot(10\ min)^{-1})$	Average diameter	SEM images
1	1	1	1	1	6.71	
2	1	2	2	2	6.93	
3	1	3	3	3	20.51	
4	2	1	2	3	4.47	
5	2	2	3	1	35.46	

Continued

TABLE 3.13 Effect of four factors on fiber average diameter and SEM figures.—cont'd

Experiment	*A* Temperature (°C)	*B* Voltage (kV)	*C* Distance (cm)	*D* MFR (g·(10 min)$^{-1}$)	Average diameter	SEM images
6	2	3	1	2	19.25	
7	3	1	3	2	24.14	
8	3	2	1	3	9.5	
9	3	3	2	1	41.72	
M_{1j}	34.15	35.32	35.46	83.89	$T = 168.69$	
M_{2j}	59.18	51.89	53.12	50.32		
M_{3j}	75.36	81.48	80.11	34.48		
R_j	41.21	46.16	44.65	49.41		

This phenomenon can be explained as follows. In the case of low viscosity, the polymer chains are free to move, some of which are deposited on the collecting plate, under the action of the electric field force, to form fibers. If the viscosity is high, a large number of chains move together, resulting in an increase in fiber thickness. Moreover, the viscosity determines the fiber diameter. Usually the temperature has the greatest effect on the viscosity of the polymer. However, when the spinning temperature is raised to a sufficiently high temperature (210–250°C, 163–175°C higher than the PP melting point), the apparent viscosity is very low. Changes in temperature within the current range do not cause significant changes in melt viscosity, so the temperature has the least effect on average diameter. On the other hand, the polymer MFR value directly reflects the melt viscosity. Its value depends mainly on the relative molecular mass and its distribution, the length of the polymer chain, the structure of the chain, the interaction between different chains, and other factors. Different MFR values mean significant differences in melt viscosity. Therefore, the MFR value is considered to be the most important factor affecting the average diameter. On the other hand, MFR is the self-factor of the material, and the voltage, distance, and temperature are external factors, and its impact on the melt cannot directly affect it.

The SEM images of the nine groups of experimental fibers are shown in Table 3.13. The fibers are compared with those obtained by solution electrospinning and the nine fibers show a smooth surface. It can be speculated that the melt electrospun fibers exhibit excellent mechanical properties. In contrast, the electrospun fibers are defective, as shown in Fig. 3.33. In Table 3.13, the minimum average diameter is 4.47 μm. The corresponding experimental conditions are A2B1C2D3, which corresponds to a temperature of 230°C, a voltage of 60 kV, a distance of 13 cm, and MFR 1000 $g\cdot(10\ min)^{-1}$.

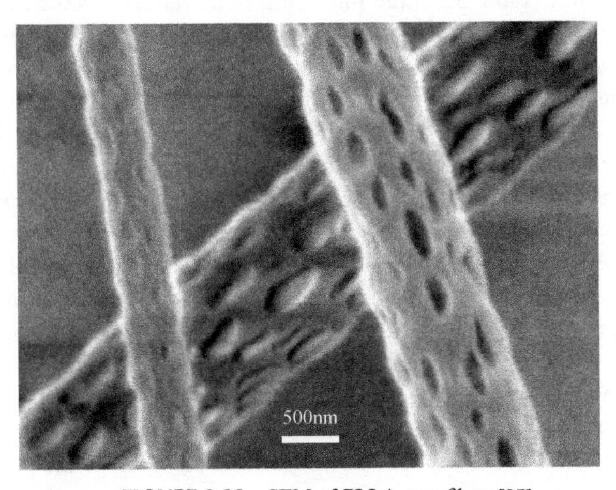

FIGURE 3.33 SEM of PLLA nanofibers [97].

TABLE 3.14 Effect on fiber diameter standard deviation [96].

Experiment	*A* Temperature (°C)	*B* Voltage (kV)	*C* Distance (cm)	*D*MFR (g·(10 min)$^{-1}$)	SD of diameter
1	1	1	1	1	0.58
2	1	2	2	2	1.24
3	1	3	3	3	1.97
4	2	1	2	3	1.03
5	2	2	3	1	7.24
6	2	3	1	2	0.52
7	3	1	3	2	5.47
8	3	2	1	3	1.20
9	3	3	2	1	14.29
M_{1j}	3.79	7.08	2.3	22.11	$T = 35.54$
M_{2j}	8.79	9.68	16.56	7.23	
M_{3j}	20.96	16.778	14.68	4.2	
R_j	17.17	9.7	14.26	17.91	

Table 3.14 shows the variance of the fiber diameter under different conditions. Variation of fiber diameter is an important feature of the fiber field. As can be seen from Table 3.14, the minimum fiber variance is 0.52 and most of the fiber variance is less than 2. This indicates that the difference in fiber diameter is small. The maximum value of R_j in the table is 17.9, indicating that the MFR has the greatest impact on average diameter and diameter distribution, which can be attributed to melt viscosity. The fiber diameter distribution order is MFR > temperature > distance > voltage.

The SEM images of nine experiments show that the melt electrospun fibers have a smoother surface compared to the solution electrospinning, which means that the melt electrospun fibers have much better mechanical properties.

In 2011, to control the spinning path, our group designed a melt electrospinning apparatus to fabricate fiber using PP as a raw material [98]. The principle of electrospinning is a strong electric field generated by applying a high voltage between the nozzle and the receiving plate. The melt is polarized in an electric field and then ejected onto the receiving plate to form ultrafine fibers. The electric field force is the only power to form the fiber, so the parallelism of the electric field has a great influence on the spinning path of the fiber.

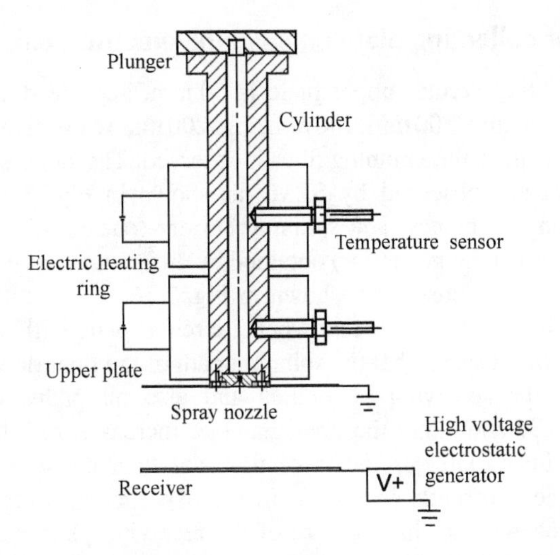

FIGURE 3.34 Schematic diagram of the electrospinning device with parallel electric field [98].

At present, in addition to the melt equipment mentioned in this chapter, all the spinning equipment cannot provide a uniform electric field for the fiber, so the spinning path control is not good, and ultrafine fiber collection is very difficult.

The equipment used in this experiment was designed by our group in the lab. See Fig. 3.34 for the parallel electric field electrospinning device diagram. The electric field is formed in a pair of parallel circular plates (receiving plate and upper plate), and both plate sizes can be adjusted. In order to protect the equipment and for safe operation, the receiver board is connected to the positive pole of the high-voltage power supply, and the upper board is grounded. In this apparatus, an efficient nozzle was used, which greatly improves the spinning efficiency and makes it easy to observe the path of the spinning fibers.

Experimental parameters: spinning temperature 230°C, spinning voltage 50 kV, receiving distance 12 cm.

First, the cylinder is heated to a preset temperature by an electric heating ring. For a variety of polymer materials, heating temperatures are closely related to their viscosity, polymer viscosity is a key factor in electrospinning [92], and therefore requires a higher temperature than the polymer molding to reduce the polymer viscosity. In addition, the temperature should not cause decomposition of the polymer. Second, after heating the cylinder to the preset temperature, the polypropylene particles are added. Finally, after the polymer melts, the nozzle produces droplets and draws fibers. Therefore, it is possible to observe the spinning fibers falling from the nozzles and being collected by the receiving plate.

3.4.2 Effect of collecting plate on spinning electric field

The diameter of the circular upper plate is 50 mm, and the diameter of the round receiver is 50 mm, 100 mm, 150 mm, and 200 mm respectively. As shown in Fig. 3.35, the path of the spinning fiber is observed. The samples obtained in each experiment are observed by SEM, as shown in Fig. 3.36. The fiber diameter, spinning efficiency, and spinning current (current displayed on the high-voltage electrostatic generator) obtained by different receiving plate areas were recorded, and the curves are shown in Fig. 3.36.

It can be seen from Fig. 3.35 that when a circular plate with a diameter of 50 mm is used, we observe that the spinning path of the fiber does not change with increasing the receiving plate area and also all paths are unstable. Figs. 3.36 and 3.37 show that the fiber diameter increases with the receiving plate area; the fiber diameter increases first and then decreases, while the spinning efficiency currently increases. In the experiment, a high voltage is applied to the receiver, so that the size of the receiving plate may affect the strength of the electric field. The larger the receiver area, the greater the electric field strength. As a result, the falling speed of the fiber becomes faster and the time required for the fiber to reach the receiver becomes shorter. So the fiber will lack sufficient stretching time, and the fiber diameter becomes thicker. However, when the falling speed of the fiber reaches a specific value, the stretching speed is sufficiently fast to completely stretch the fiber to obtain a smaller fiber diameter.

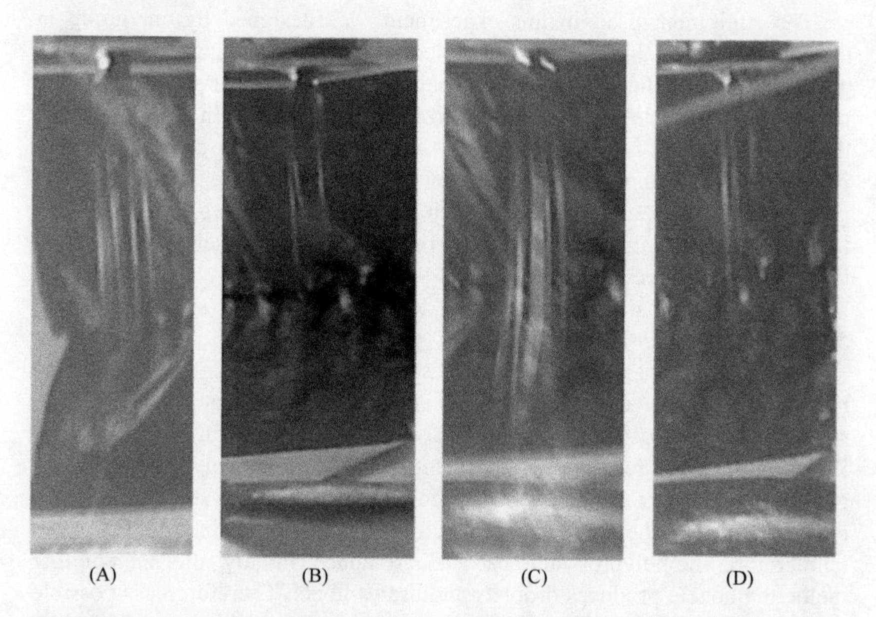

(A) (B) (C) (D)

FIGURE 3.35 Photograph of the fibers' spinning path produced using an upper plate diameter of 50 mm, and a receiver diameter of (A) 50 mm, (B) 100 mm, (C) 150 mm, and (D) 200 mm [98].

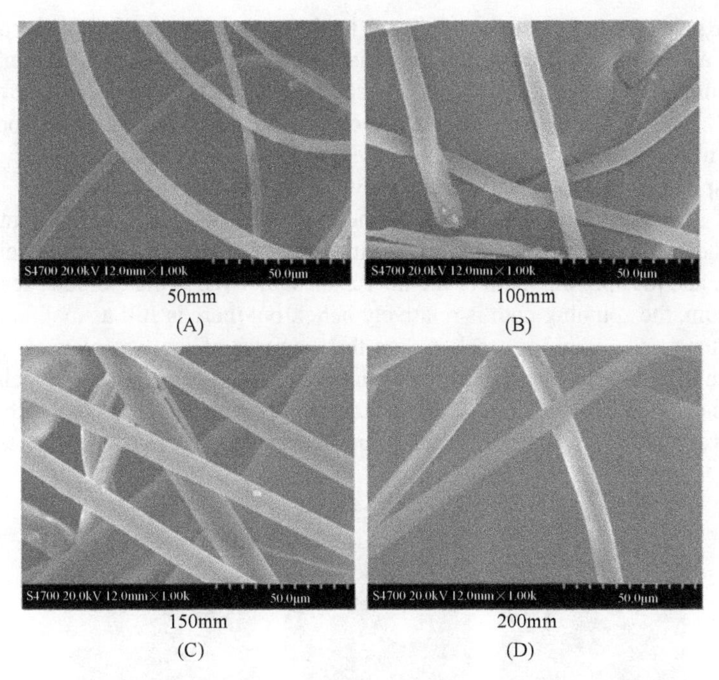

FIGURE 3.36 SEM images of PP melt electrospinning fibers produced using an upper plate diameter of 50 mm, and a receiver plate diameter of (A) 50 mm, (B) 100 mm, (C) 150 mm, and (D) 200 mm, respectively [98].

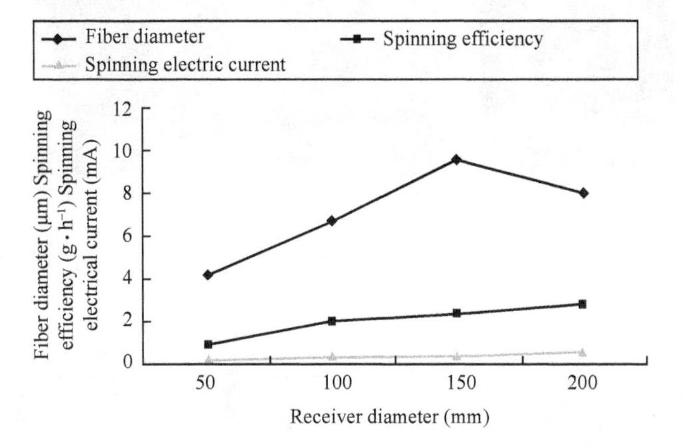

FIGURE 3.37 Variation of average fiber diameter, spinning efficiency, and spinning electrical current [98].

3.4.3 Effect of upper plate on spinning electric field

It was found that the area of the upper plate had little effect on the fiber spinning path and the spinning electric field.

A pair of parallel plates was used in the experiment, including an upper plate and a receiver having the same diameter, 50 mm, 100 mm, 150 mm, and 200 mm, respectively. The path of the spinning fibers is observed, as shown in Fig. 3.38. The samples obtained for each test are observed by SEM and are shown in Fig. 3.39. Variations of average fiber diameter, spinning efficiency, and spinning electrical current are shown in Fig. 3.40.

It is observed that when the diameters of the two plates are 50 mm, the distance between them is too large with respect to the area of the receiving plate, and the spinning path of the fiber is unstable. When the diameter reaches 100 mm, the spinning path is relatively better, but there is still a small amount of jitter on the spinning track. When the diameter of the two plates became 150 mm, it is observed that the spinning path of the fibers is almost vertical. When the diameter is as large as 200 mm, the spinning path was also vertical but, in addition, there is an increase in the area of the receiving

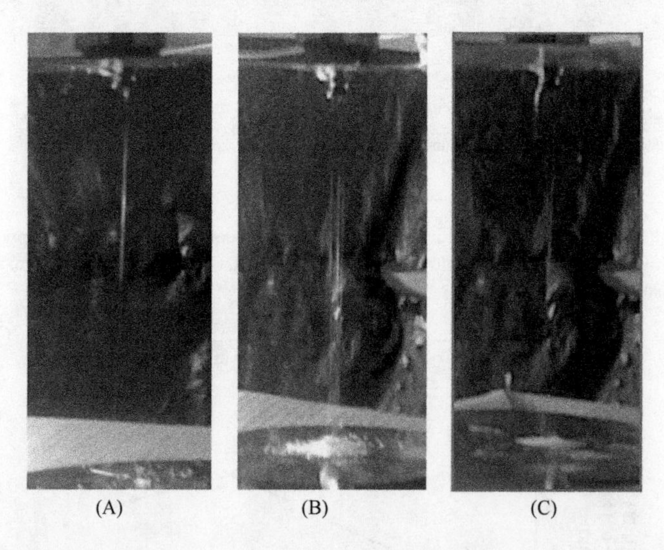

(A) (B) (C)

FIGURE 3.38 Photograph of the fibers' spinning path produced using both parallel plates with diameters of (A) 100 mm, (B) 150 mm, and (C) 200 mm [98].

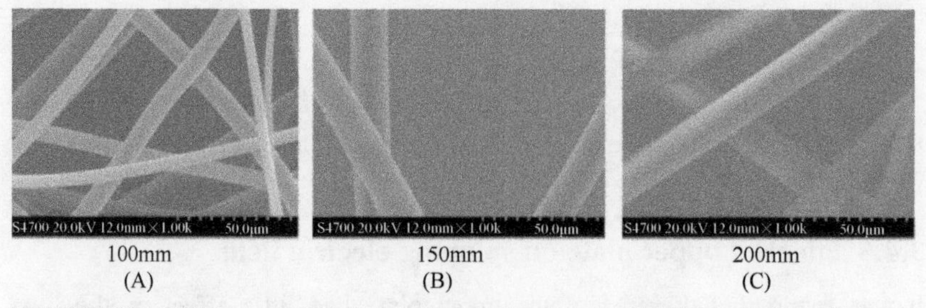

100mm (A) 150mm (B) 200mm (C)

FIGURE 3.39 SEM images of PP melt electrospinning fibers produced using both parallel plates with diameters of (A) 100 mm, (B) 150 mm, and (C) 200 mm [98].

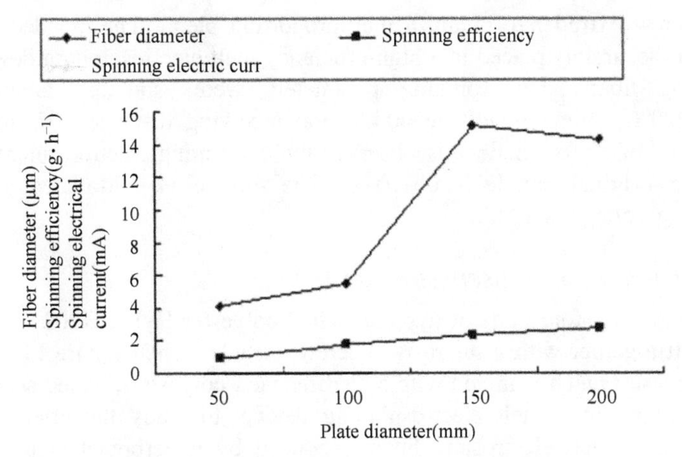

FIGURE 3.40 Variation of average fiber diameter, spinning efficiency, and spinning electrical current [98].

plate, spinning efficiency, and current. Fig. 3.40 shows the variations in fiber diameter. From the above analysis we can see that the receiver area on the spinning electric field strength, spinning current, and spinning efficiency has a great impact, when the receiver area increases. Through a series of experiments, it was found that a pair of parallel plates of the same size contributes to the formation of parallel electric fields. When the diameter between the two plates is greater than the distance between them, the spinning path becomes vertical to form a structured electric field.

3.4.4 Effect of the hyperbranched polymers

In 2012, Xia and our group revealed [99] the use of different molecular weights, different contents of hyperbranched polyester on the modified polypropylene, and the effect of hyperbranched polymer on the diameter of polypropylene melt electrospinning fiber.

Hyperbranched polymers are a class of macromolecules with highly branched and three-dimensional quasispherical stereostructures [100−103]. They have unique physical and chemical properties, such as low-melt viscosity [104,105], high rheology [106], good solubility [107], a large number of intramolecular holes, and modifiable terminal functional groups. In the polymer processing and modification they have broad application prospects [108]. We have explored hyperbranched polymers in melt electrospinning. The following describes the specific experimental process.

3.4.4.1 Sample preparation

The refining hyperbranched polymer is mixed with polypropylene in different proportions followed by blending and extruding at 180−200°C with a micro twin-screw extruder to obtain a multiproportion blend. The blended master batch is separately placed in a high-efficiency melt electrospinning device for spinning fibers. The spinning parameters were: spinning temperature 200–220°C, spinning voltage 40 kV, and receiving distance 105 mm. The diameter of the spun fiber is observed by a scanning electron microscope, and the original sample is observed. The surface is gold-plated and the scanning voltage is 20 kV.

3.4.4.2 Result and discussion

Firstly, we add four parts of hyperbranched polyester HyPer H202 to PP and blend it together with a micro twin-screw extruder. Then a pure PP and PP blend master batch is mixed with hyperbranched polyesters, added separately for spinning in a melt electrospinning device, to study the effect of the diameter of melt electrospun fibers produced by hyperbranched polyesters mixed with PP. Fig. 3.41A shows a schematic diagram of a melt electrospun PP single fiber, fabricated at the following parameters: spinning temperature 200–220°C, spinning voltage 40 kV, and receiving distance 105 mm. Under the process conditions, the PP melt electrospun fiber diameter can reach 5 μm; Fig. 3.41B shows a melt spinning H202/PP single-fiber schematic. Adding four parts to PP HyPer H202 reduces the fiber diameter to 20 nm.

Therefore, this reduction in diameter is mainly due to the three-dimensional quasispherical structure of HyPer H202 penetrated into the linear PP polymer chain, and thus results in reducing the entanglement between the molecular chains. On the other hand, the viscosity of PP molecules is relatively low, which plays a role in the slide, to promote the molecular chain slip and results in a great reduction of PP melt viscosity. Under the action of high voltage, the low viscosity of melt can easily overcome the surface tension to form a fine jet, and is sprayed. Curing takes place during the process and the jet eventually falls onto the receiving device to form micronanofibers [110,111].

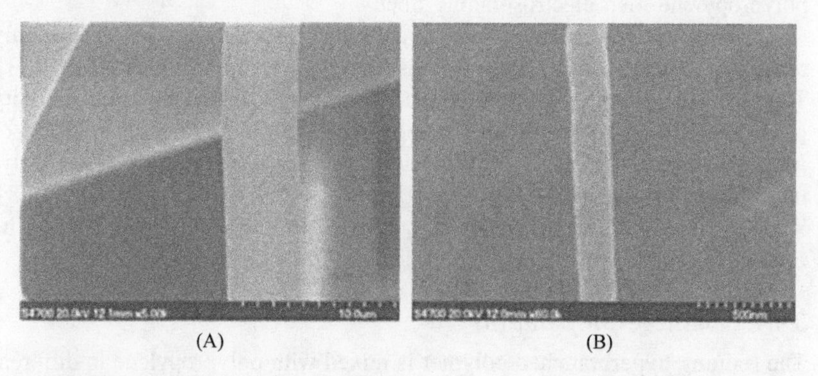

(A) (B)

FIGURE 3.41 SEM images of (A) PP single fiber, (B) H202/PP single fiber [109].

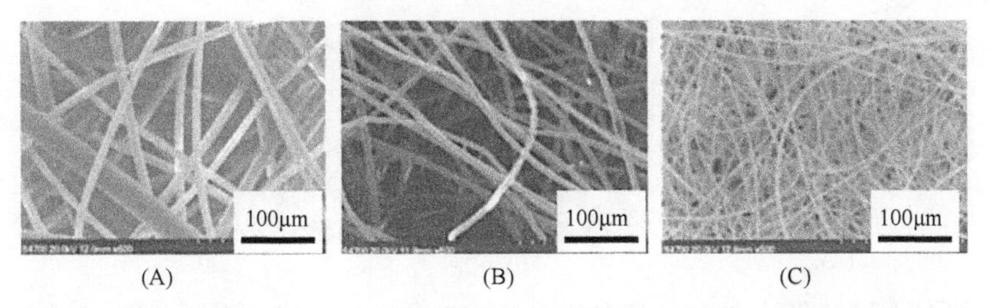

FIGURE 3.42 SEM images of (A) PP fibers, (B) H202/PP fibers, and (C) H203/PP fibers [109].

HyPer H202 and H203 are based on heat-resistant polyester hyperbranched polymers. They have high molecular weight, highly active terminal hydroxyl hyperbranched polymers, and their hydroxyl groups and molecular weights are different. The molecular weight of H202 is about 2500 and H203 is about 5500. By taking the same mass fraction of HyPer H202 and H203 added to PP, they are granulated by a microextruder followed by spinning in a melt electrospinning device.

The effect of the molecular weight of hyperbranched polyesters on melt electrospinning was studied. Experimental results are shown in Fig. 3.42.

The above figure shows that terminal hydroxyl groups are similar in structure but have different molecular weights. Hyperbranched polymers have significant effects on the diameter of melt electrospun PP fibers. PP fibers produced by melt electrospinning have a nonuniform diameter distribution of 5−20 μm.

H202-modified PP has a fiber diameter of 4−5 μm and H203-modified PP has a fiber diameter of 1−2 μm. This difference between them in fiber diameter is mainly because H203 is more moderate in molecular weight than H202 and can more easily penetrate into the linear PP polymer chain, thus resulting in reducing the entanglement between the molecular chains. Moreover, H203-modified PP melts have low melt viscosity. Therefore, on the application of high voltage, the low viscosity of melt can easily overcome the surface tension to form a fine jet and produce fibers with a small diameter and uniform distribution.

PP was added in 0, 4, 8, and 16 parts of the hyperbranched polyester HyPer H203, extruded in a microextruder granulation, and then into the high-performance melt electrospinning machine for spinning. The effect of the hyperbranched polyester content on the diameter of the melt spinning PP fiber was investigated, as shown in Figs. 3.42A and 3.43. According to Fig. 3.43, it was found that the diameter of the PP melt electrospinning fibers without hyperbranched polyester was not uniform and was 5−20 μm. When the

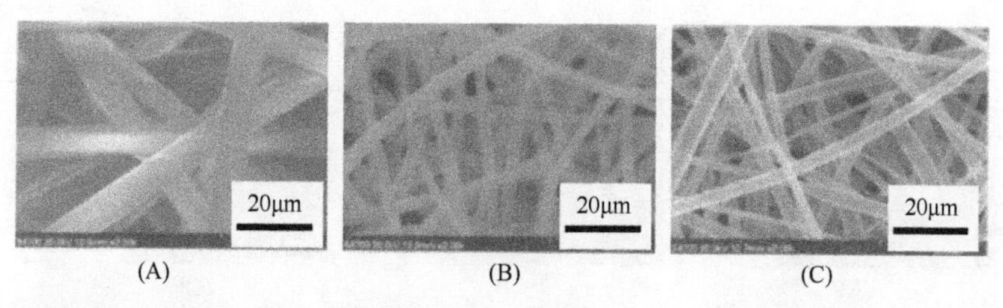

FIGURE 3.43 SEM images of PP fibers containing different contents of H203 in PP: (A) four copies; (B) eight copies; and (C) 16 copies [109].

content of hyperbranched polyester H203 is 4 % (*w*), the diameter of the PP fiber is 5–6 μm; with eight copies, the preparation of PP fiber diameter is 1–2 μm; with 16 copies, the fiber diameter distribution is not uniform, mostly below 4 μm, but part of the fiber diameter can reach 200 nm. This is because, when the hyperbranched polyester H203 penetrates into the entangled linear PP polymer chain, the content is a relatively low part of the PP polymer chain to be entangled. With the increase in H203 content, more and more PP polymer chains were entangled, and under the lubrication of H203, the slippage between the PP chains was further promoted, and the melt viscosity of PP became lower and lower. In the high-voltage electrostatic field, the fiber diameter is greater and finer, so when the content of hyperbranched polyester H203 is 8 % (*w*), the PP fiber diameter is lower than the hyperbranched polyester content of 0 parts, four copies of PP fiber. However, when the content of hyperbranched polyester H203 is 16 % (*w*), the fiber diameter distribution is no longer uniform, the fiber diameter is mostly below 4 μm, and hundreds of nanometer fibers are formed, and some fiber diameters can reach 200 nm. This is because the hyperbranched polyester content of 16 parts of PP melt viscosity is very low. In the existing spinning temperature, spinning voltage, and receiving distance process conditions, most of the PP melt is too late in the high-voltage electrostatic field, and has been fully stretched to the receiving device, so that fiber diameter is not a uniform phenomenon.

(1) PP fiber electrospinning single-fiber diameter can reach 5 μm, adding four copies of hyperbranched polyester H202, and the single fiber diameter can be reduced to 200 nm.

(2) The effect of hyperbranched polyester with different molecular weights on the diameter of PP fiber was significantly different. The average diameter of H203-modified PP fiber was about 3 μm lower than that of H202.

(3) The PP melt electrospinning fibers without hyperbranched polyester have a nonuniform diameter distribution of 5–20 μm. With the increase in hyperbranched polyester content, PP melt electrospun fiber diameter decreases. When adding eight parts of hyperbranched polyester H203, the fiber diameter can reach 1–2 μm, the distribution is uniform, the hyperbranched polyester content is too high, and the fiber diameter becomes uneven.

3.4.5 Effect of polar additive on PP

Liu et al. [112] also studied the effect of polar additives on the electrospinning of nonpolar polypropylene melt. Polypropylene (PP) is not easily soluble at an ambient temperature. Therefore, electrospinning is usually carried out in the molten state. However, the melt electrospun PP fibers are more nonuniform and have a larger diameter than the solution electrospun fibers. This is mainly because the melt viscosity is too large and, in order to reduce the viscosity and get a smaller diameter, manufacture of fiber has been attempted in two ways. First, the conditions for melt spinning [111−115] can be changed, such as voltage, melt temperature, distance of reception, and the like. Second, the polymer content may be changed, for example, by adding a plasticizer [111] or a polymer [90] for reducing the molecular weight. Malakhov et al. [98] used sodium stearate and oleate as an additive to reduce the viscosity of the polyamide 6 melt in electrospinning. They found that adding 10 %(w) of the additive reduced the average fiber diameter by 40 times due to a 60% reduction in melt viscosity. Polyamide 6, sodium stearate, and operate are polar materials, and there is a strong interaction between these molecules, therefore the use of polar additives is effective. However, nonpolar polymers such as PP, polyethylene, polybutadiene, and polystyrene are very common in the fiber industry. Therefore, it should be determined whether the polar additive also affects the nonpolar polymer in melt electrospinning. In this study, in order to solve this problem, the polar additives, namely stearic acid and sodium stearate, were added to pure PP. The effects of additives were studied. The results show that the fiber diameter of PP is 8%(w), and the stearic acid is reduced by 69.3% (5.4−1.6 μm) compared with pure PP. When the sodium stearate content is 10 and 8%(w), the minimum fiber diameter is 600 nm and the minimum average fiber diameter is 1.8 m. The addition of polar compounds not only changes the diameter of the PP microfibers but also changes the diameter distribution, the processing current, and even the thermal properties of the fibers. We have experimented with the microscopic mechanisms of these changes.

3.4.5.1 Material

PP (iPP) was obtained from Shanghai Expert in the Developing of New Material Co., China. The melt flow index of the PP was 1500 g·(10 min)$^{-1}$. Stearic acid and sodium stearate were purchased from Xilong Chemical Co., China. Stearic acid, with a relative molecular mass of 284.48, is a short-chain polymer. Sodium stearate, with a relative molecular mass of 306.46, is an organic salt. All reagents are used as received without any further treatment.

3.4.5.2 Effect of polar additive on electrospinning current

Polar additives are easily polarized under strong electric fields. Therefore, the addition of polar material to PP will result in a fundamental change in the

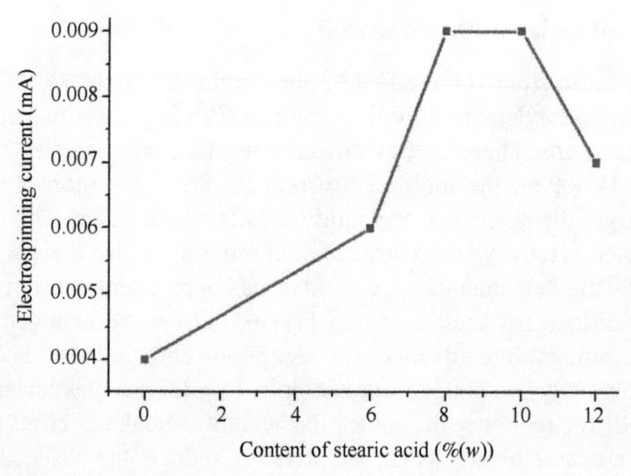

FIGURE 3.44 Changes in the electrospinning current of PP containing different contents of stearic acid [111].

electrospinning process. During the course of the experiment, the current of the high-voltage generator is recorded when PP or its composite material with different stearic acid content is spun. Under the same spinning conditions, the current increases with an increase in stearic acid content, indicating that the polarized electron cloud of stearic acid increases the charge of the fiber containing the additive (Fig. 3.44). However, when the stearic acid content was increased from 10 to 12 %(w), the current decreased. Fig. 3.45 explains this phenomenon. Although the amount of fine stearic acid particles in the surface layer of the crude fiber is equal to or slightly higher than that of the thin fiber surface layer, the charge density of the thicker fiber surface area is low. In addition, when the stearic acid content reaches 12 %(w), more stearic acid molecules may aggregate to form larger domains, while fewer molecules are dispersed in the PP melt phase. Stearic acid cannot be evenly distributed in the PP phase, which results in a discontinuous distribution of fibers that reduces the number of charges in the cell. Thus, when the stearic acid is 12.%(w), the current is lower than the content of 10 %(w).

3.4.5.3 Effect of polar additives on fiber diameter

The polar additive significantly affects the fiber diameter of PP. SEM images of PP with different amounts of polar additives are shown in Fig. 3.46, and the statistical data are shown in Table 3.15. With the increase in stearic acid or sodium stearate content, the average fiber diameter of PP composites was first reduced (Fig. 3.46 and Table 3.15). When the stearic acid content is 8 %(w) and the minimum diameter is 1.65 m, the SD is not the smallest. The minimum SD appears in 10 %(w) stearic acid. However, when the content exceeds 10 % (w), the average diameter increases.

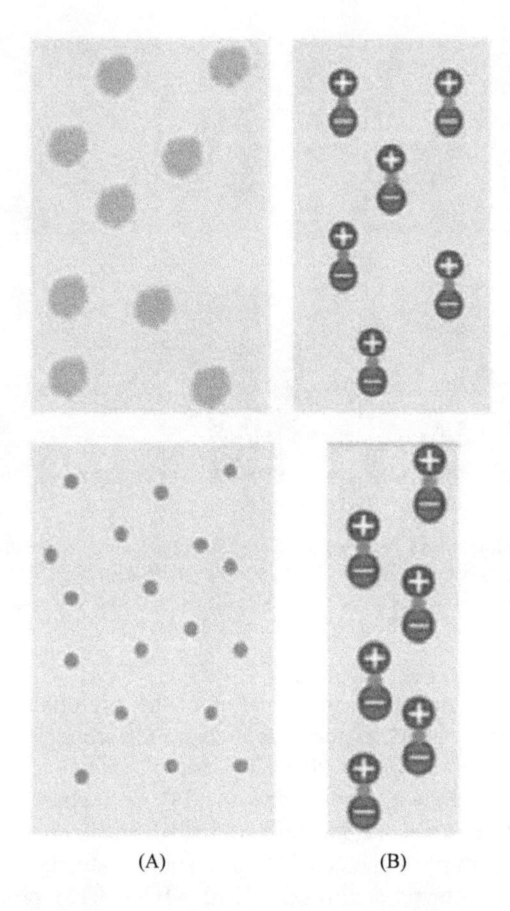

FIGURE 3.45 Mimic diagrams of different PP fibers containing additives. The *dark rectangles* represent fibers. (A) The PP fibers with sodium stearate (irregular particles) (upper) and with stearic acid (points) (lower). (B) The PP fibers with stearic acid (dipole) electrospun at low (upper) and high (lower) voltages [111].

The possible cause of the change in diameter is the presence of carboxyl groups in the stearic acid molecule. In the high electric field, the carboxyl oxygen double bond electron cloud and carboxyl oxygen single bond are easy to change and polarize. Then, the connection of the polymer chains to stearic acid, entanglement, or encapsulation will withstand strong electric field forces. In addition, stearic acid can be plasticized PP. With the increase in stearic acid content, the PP composite melt is affected by the electric field. At high pulling forces, the diameter of the PP microfibers formed at the end of the Taylor cone is small. However, when the stearic acid content was increased to $12\%\,(w)$, the fiber diameter became thick again, at 4.8 m.

FIGURE 3.46 SEM images of PP fibers containing different contents of additive, electrospun at 220°C, 10 cm spinning distance, and 30 kV: (A) 6%(w) stearic acid; (B) 6%(w) sodium stearate; (C) 8%(w) stearic acid; (D) 8%(w) sodium stearate; (E) 10%(w) stearic acid; (F) 10%(w) sodium stearate; (G) 12% (w) stearic acid; and (H) without additives [111].

This may be due to the increased formation of the microphase of the stearic acid molecule, and when the stearic acid content reaches 12% (w), the molecular dispersion is low in the PP melt phase. Thus, the continuous stearic acid phase is very easy to move, and the PP melt phase becomes more viscous, resulting in a second increase in fiber diameter.

The minimum fiber diameter of PP and sodium stearate is significantly smaller than that of stearic acid (Fig. 3.47 and Table 3.15), probably because sodium stearate is a salt. The salt can be easily ionized in the electric field. Positive and negative ions move to the opposite end, and then generate electricity in the direction of the electric field. For stearic acid under strong electric fields, only the migration of the electron cloud occurs and the polarization is much smaller than that obtained with sodium stearate. The force of electron movement is significantly greater than the force of electron cloud movement. Thus, the fiber diameter of the PP composite containing sodium stearate is smaller than the fiber diameter of the PP composite containing stearic acid. However, the diameter distribution of PP and sodium stearate is very uneven compared to sodium stearate. This may be because sodium stearate is a hydrophilic salt that does not have plasticization of the organic PP melt. Stearic acid as an organic material can be uniformly distributed in the PP melt and has a significant plasticizing effect on the melt. This phenomenon leads to the presence of thick and thin fibers in PP and sodium stearate, resulting in uneven diameters.

TABLE 3.15 Statistical data of diameters of PP fibers containing different additive contents [111].

Parameter	6% (*w*) stearic acid	6% (*w*) sodium stearate	8% (*w*) stearic acid	8% (*w*) sodium stearate	10% (*w*) stearic acid	10% (*w*) sodium stearate	1.2% (*w*) stearic acid	Without additives
AVG/μm	2.1	2.4	1.6	1.8	2.1	2.2	4.9	5.4
SDEV	0.2	0.7	0.1	0.8	0.1	2.2	0.2	1.7
MIN/μm	1.7	0.9	1.3	0.8	2.0	0.6	4.5	3.0

AVG, average diameter; MIN, minimum value; SDEV, standard values of deviation.

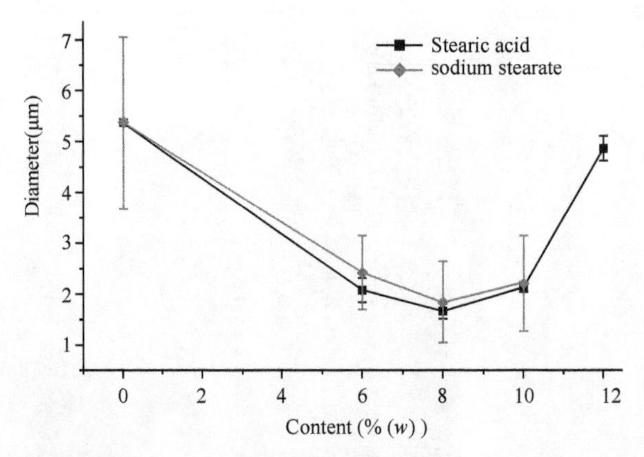

FIGURE 3.47 Changes in the average diameter of PP fibers containing additives at different contents of stearic acid and sodium stearate [111].

3.5 Conclusion

This chapter introduces our research using self-designed melt electrospinning equipment to prepare nanofibers with different materials, and analyzes the morphological and mechanical properties of the nanofibers by various characterizations. The influence factors are compared and analyzed by orthogonal experiment method, and the order of influencing factors is obtained. Finally, the optimum spinning conditions are obtained with fine and uniform average fiber diameter. This has laid a good scientific theoretical basis for future experiments.

References

[1] Hu XY, Gao ZY, Liu HF, et al. Research progress on the modification of biodegradable plastic polycaprolactone. Engineering Plastics Applications 2015;43(11):108–11.

[2] Pappalardo D, Annunziata L, Pellecchia C, et al. Ring-opening polymerization of ε-caprolactone by benzylalkoxybis (2,4,6-triisopropylphenyl) tin compounds: observation of the insertion product into the Sn–OMe bond. Macromolecules 2007;6(40):1886–90.

[3] Romeo V, et al. Encapsulation and exfoliation of inorganic lamellar fillers into polycaprolactone by electrospinning. Biomacrolecules 2007;8(10):3147–52.

[4] Liu Y, Deng RJ, Qi HX, et al. Experimental study on melt electrospinning of polycaprolactone. In: National symposium on Polymer Science; 2009 [Tianjin, China].

[5] Yang WM, Deng RJ, Ding YM. A magnetic field assisted polymer melt electrospinning device, vol. 8; 2009.

[6] Yang WM, Deng RJ, Liu Y, et al. A highly efficient electrospinning nozzle, vol. 7; 2009.

[7] Gupta B, Revagade N, Hilborn J. Poly (lactic acid) fiber: an overview. Progress in Polymer Science 2007;32(4):455–82.

[8] Engelberg I, Kohn J. Physico-mechanical properties of degradable polymers used in medical applications: a comparative study. Biomaterials 1991;12(3):292–304.

[9] Tong H, Hu J, Shen X, et al. Preparation and characterization of homogeneous chitosan–polylactic acid/hydroxyapatite nanocomposite for bone tissue engineering and evaluation of its mechanical properties. Acta Biomaterialia 2009;5(7):2693–703.

[10] Wee YJ, Kim JN, Yun JS, et al. Utilization of sugar molasses for economical l(+)-lactic acid production by batch fermentation of *Enterococcus faecalis*. Enzyme and Microbial Technology 2004;35(6–7):568–73.

[11] Benny O, Fainaru O, Adini A, et al. An orally delivered small-molecule formulation with antiangiogenic and anticancer activity. Nature Biotechnology 2008;26(7):799–807.

[12] Conn RE, Kolstad JJ, Borzelleca JF, et al. Safety assessment of polylactide (PLA) for use as a food-contact polymer. Food and Chemical Toxicology 1995;33(4):273–83.

[13] Glenn G, Klamczynski A, Ludvik C, et al. In situ lamination of starch-based baked foam packaging with degradable films. Packaging Technology and Science 2010;20(2):77–85.

[14] Iovino R, Zullo R, Rao MA, et al. Biodegradation of poly (lactic acid)/starch/coir biocomposites under controlled composting conditions. Polymer Degradation and Stability 2008;93(1):147–57.

[15] Rychter P, Biczak R, Herman B, et al. Environmental degradation of polyester blends containing atactic poly (3-hydroxybutyrate). Biodegradation in soil and ecotoxicological impact. Biomacromolecules 2006;7(11):3125–31.

[16] Hunley MT, Long TE. Electrospinning functional nanoscale fibers: a perspective for the future. Polymer International 2008;57(3):385–9.

[17] McCullen SD, Stano KL, Stevens DR, et al. Development, optimization, and characterization of electrospun poly(lactic acid) nanofibers containing multi-walled carbon nanotubes. Journal of Applied Polymer Science 2007;105(3):1668–78.

[18] Cho AR, Shin DM, Jung HW, et al. Effect of annealing on the crystallization and properties of electrospun polylactic acid and nylon 6 fibers. Journal of Applied Polymer Science 2011;120(2):752–8.

[19] Mizutani Y, Hattori M, Okuyama M, et al. Poly(l-lactic acid) short fibers prepared by solvent evaporation using sodium tripolyphosphate. Polymer 2005;46(11):3789–94.

[20] Xu AC, Zhao JN, Pan ZJ, et al. The microstructure and mechanical behavior of electrospinning polylactic acid fiber felt. Journal of Textile Research 2007;28(7):4–8.

[21] Zhao ML, Sui G, Deng XL, et al. Spinning of polylactic acid nanofibers by electrospinning. Synthetic Fiber Industry 2006;29(1):5–7.

[22] Ge PF. Preparation and structural properties of polylactic acid nanofiber, vol. 46. Jiangnan University; 2007.

[23] Mai HZ, Zhao YM, Nie FM. Research progress in spinning of biodegradable polylactic acid fibers. Synthetic Fiber Industry 2000;23(4):43–5. 49.

[24] Liu Y, Ding Y M, Yan H, et al. Research on melt electrospinning of polylactic acid, seminar on the 10th functional textile and nanotechnology application of Xuelian Cup 2010, vol. 4. [Changzhou, Jiangsu, China].

[25] Deng R, Liu Y, Ding Y, et al. Melt electrospinning of low-density polyethylene having a low-melt flow index 2009;114(1):166–75.

[26] Yang WM, Deng RJ, Liu Y, et al. High efficiency electrospinning nozzle. 2009. CN 101570898 A.

[27] Zhao FW, Liu Y, Ding YM, et al. Effect of plasticizer and load on melt electrospinning of PLA. Key Engineering Materials 2012;501:32–6.

[28] Gai X, Yan W, Han X, et al. Pulsed electric fields on poly-L-(lactic acid) melt electrospun fibers. Industrial and Engineering Chemistry Research 2016;55(26).

[29] Li X, Liu Y, Peng H, et al. Effects of hot airflow on macromolecular orientation and crystallinity of melt electrospun poly (L-lactic acid) fibers. Materials Letters 2016;176:194–8.
[30] Liu Y, Li X, Ramakrishna S. Melt electrospinning in a parallel electric field. Journal of Polymer Science Part B: Polymer Physics 2014;52(14):946–52.
[31] Brown TD, Dalton PD, Hutmacher DW. Melt electrospinning today: an opportune time for an emerging polymer process. Progress in Polymer Science 2016;56:116–66.
[32] Richardlacroix M, Pellerin C. Partial disentanglement in continuous polystyrene electrospun fibers. Macromolecules 2015;48(1):37–42.
[33] Ghashghaie S, Bazargan AM, Ganji ME, et al. An investigation on the behavior of electrospun ZnO nanofibers under the application of low frequency AC electric fields. Journal of Materials Science: Materials in Electronics 2011;22(9):1303–7.
[34] Li JL. EHD sprayings induced by the pulsed voltage superimposed to a bias voltage. Journal of Electrostatics 2007;65(12):750–7.
[35] Zhao F, Liu Y, Yuan H, et al. Orthogonal design study on factors affecting the degradation of polylactic acid fibers of melt electrospinning. Journal of Applied Polymer Science 2012;125(4):2652–8.
[36] Liu Y, Zhao FW, Zhang C, et al. Solvent-free preparation of poly(lactic acid) fibers by melt electrospinning using an umbrella-like spray head and alleviation of the problematic thermal degradation. Journal of the Serbian Chemical Society 2012;77(8):1071–82.
[37] Lyons J, Li C, Ko F. Melt-electrospinning part I: processing parameters and geometric properties. Polymer 2004;45(22):7597–603.
[38] Cao J, Covarrubias VM, Straubinger RM, et al. A rapid, reproducible, on-the-fly orthogonal array optimization method for targeted protein quantification by LC/MS and its application for accurate and sensitive quantification of carbonyl reductases in human liver. Analytical Chemistry 2010;82(7):2680–9.
[39] Qian J, Yang Q, Sun F, et al. Cogeneration of biodiesel and nontoxic rapeseed meal from rapeseed through in-situ alkaline transesterification. Bioresource Technology 2013;128:8–13.
[40] Tao W, Xungang D, Peng D. Orthogonal optimization for room temperature magnetron sputtering of ZnO:Al films for all-solid electrochromic devices. Applied Surface Science 2011;257(8):3748–52.
[41] Xie G, Chen ZY, Ramakrishna S, et al. Orthogonal design preparation of phenolic fiber by melt electrospinning. Journal of Applied Polymer Science 2015;132(38):1–8.
[42] Fong H, Chun I, Reneker DH. Beaded nanofibers formed during electrospinning. Polymer 1999;40(16):4585–92.
[43] Zhou H, Green TB, Joo YL. The thermal effects on electrospinning of polylactic acid melts. Polymer 2006;47(21):7497–505.
[44] Garlotta D. A literature review of poly(lactic acid). Journal of Polymers and the Environment 2001;9(2):63–84.
[45] Zhang J, Duan YX, Sato H, et al. Crystal modifications and thermal behavior of poly(l -lactic acid) revealed by infrared spectroscopy. Macromolecules 2005;38(19):8012–21.
[46] Zong XH. Structure and process relationship of electrospun bioabsorbable nanofiber membranes. Polymer 2002;43(16):4403–12.
[47] Monticelli O, Bocchini S, Gardella L, et al. Impact of synthetic talc on PLLA electrospun fibers. European Polymer Journal 2013;49(9):2572–83.

[48] Ping S, Guangyi C, Zhiyong W, et al. Calorimetric analysis of the multiple melting behavior of melt-crystallized poly(l-lactic acid) with a low optical purity. Journal of Thermal Analysis and Calorimetry 2013;111(2):1507−14.
[49] Bas O, De JP, Elena M, et al. Enhancing structural integrity of hydrogels by using highly organised melt electrospun fibre constructs. European Polymer Journal 2015;72:451−63.
[50] Cho D, Zhmayev E, Joo YL. Structural studies of electrospun nylon 6 fibers from solution and melt. Polymer 2011;52(20):4600−9.
[51] Inai R, Kotaki M, Ramakrishna S. Structure and properties of electrospun PLLA single nanofibres. Nanotechnology 2005;16(2):208−13.
[52] Liu L, Ren Y, Li Y, et al. Effects of hard and soft components on the structure formation, crystallization behavior and mechanical properties of electrospun poly(l-lactic acid) nanofibers. Polymer 2013;54(19):5250−6.
[53] Yu L, Liu HS, Dean K, et al. Cold crystallization and postmelting crystallization of PLA plasticized by compressed carbon dioxide. Journal of Polymer Science Part B: Polymer Physics 2010;46(23):2630−6.
[54] Nisha SK, Asha SK. Random copolyesters containing perylene bisimide: flexible films and fluorescent fibers. ACS Applied Materials and Interfaces 2014;6(15):12457−66.
[55] Yano S, Kurita K, Iwata K, et al. Structure and properties of poly(vinyl alcohol)/tungsten trioxide hybrids. Polymer 2003;44(12):3515−22.
[56] Zhang J, Tashiro K, Domb AJ, et al. Confirmation of disorder α form of poly(L-lactic acid) by the X-ray fiber pattern and polarized IR/Raman spectra measured for uniaxially-oriented samples. Macromolecular Symposia 2010;242(1):274−8.
[57] Tanaka MYR. Molecular orientation distributions in uniaxially oriented poly(L-lactic acid) films determined by polarized Raman spectroscopy. Biomacromolecules 2006;7(9):2575.
[58] Ribeiro C, Sencadas V, Costa CM, et al. Tailoring the morphology and crystallinity of poly(L-lactide acid) electrospun membranes. Science and Technology of Advanced Materials 2011;12(1):015001.
[59] Park MS, Wong YS, Park JO, et al. A simple method for obtaining the information of orientation distribution using polarized Raman spectroscopy: orientation study of structural units in poly (lactic acid). Macromolecules 2011;44(7):2120−31.
[60] Yu Y, Wang Y, Lin K, et al. Complete Raman spectral assignment of methanol in the C−H stretching region. The Journal of Physical Chemistry A 2013;117(21):4377−84.
[61] Qu J, Zhang X, Jin G. Orientation kinetics of screw-axial vibration on glass fiber reinforced polypropylene composites. Journal of Macromolecular Science Part D − Reviews in Polymer Processing 2008;47(2):13.
[62] Greenfeld I, Zussman E. Polymer entanglement loss in extensional flow: evidence from electrospun short nanofibers. Journal of Polymer Science Part B: Polymer Physics 2013;51(18):1377−91.
[63] Zussman E, Rittel D, Yarin AL. Failure modes of electrospun nanofibers. Applied Physics Letters 2003;82(22):3958−60.
[64] Taiyo Y, Roland D, Andreas G, et al. Highly oriented crystalline PE nanofibrils produced by electric-field-induced stretching of electrospun wet fibers. Macromolecular Materials and Engineering 2010;295(12):1082−9.
[65] Gaudio D, Ercolani C, et al. Assessment of poly (e-caprolactone)/poly(3-hydroxybutyrate-co-3-hydroxyvalerate) blends processed by solvent casting and electrospinning. Materials Science and Engineering A 2011;528(3):1764−72.
[66] Mckee MG, Layman JM, Cashion MP, et al. Phospholipid nonwoven electrospun membranes. Science 2006;311(5759):353−5.

[67] Economy J, Clark RA. Fibers from novolacs. 1972.

[68] Gugumus F. New trends in the stabilization of polyolefin fibers. Polymer Degradation and Stability 1994;44(3):273–97.

[69] Wang MX, Huang ZH, Kang F, et al. Porous carbon nanofibers with narrow pore size distribution from electrospun phenolic resins. Materials Letters 2011;65(12):1875–7.

[70] Wang L, Huang ZH, Yue MB, et al. Preparation of flexible phenolic resin-based porous carbon fabrics by electrospinning. Chemical Engineering Journal 2013;218(3):232–7.

[71] Nair CPR, Bindu RL, Ninan KN. Thermal characteristics of addition-cure phenolic resins. Polymer Degradation and Stability 2001;73(2):251–7.

[72] Liu CL, Ying YG, Feng HL, et al. Microwave promoted rapid curing reaction of phenolic fibers. Polymer Degradation and Stability 2008;93(2):507–12.

[73] Zhang D, Shi J, Guo Q, et al. Preparation mechanism and characterization of a novel, regulable hollow phenolic fiber. Journal of Applied Polymer Science 2007;104(4):2108–12.

[74] Zhang CL, Yu SH. Nanoparticles meet electrospinning: recent advances and future prospects. Chemical Society Reviews 2014;43(13):4423.

[75] Pavliňák D, Hnilica J, Quade A, et al. Functionalization and pore size control of electrospun PA6 nanofibres using microwave jet plasma. Polymer Degradation and Stability 2014;108:48–55.

[76] Kancheva M, Toncheva A, Manolova N, et al. Advanced centrifugal electrospinning setup. Materials Letters 2014;136(136):150–2.

[77] Teng M, Qiao J, Li F, et al. Electrospun mesoporous carbon nanofibers produced from phenolic resin and their use in the adsorption of large dye molecules. Carbon 2012;50(8):2877–86.

[78] Bai Y, Huang ZH, Kang F. Electrospun preparation of microporous carbon ultrafine fibers with tuned diameter, pore structure and hydrophobicity from phenolic resin. Carbon 2014;66:705–12.

[79] Wang M, Wu YN, Shen JY, et al. Preparation, characterization and electrochemical measurement of porous carbon derived from poly(furfuryl alcohol)/polyvinylpyrrolidone electrospun nanofibers. RSC Advances 2014;4(108):63162–70.

[80] Imaizumi S, Matsumoto H, Suzuki K, et al. Phenolic resin-based carbon thin fibers prepared by electrospinning: additive effects of poly(vinyl butyral) and electrolytes. Polymer Journal 2009;41(12):1124–8.

[81] Ma C, Song Y, Shi J, et al. Phenolic-based carbon nanofiber webs prepared by electrospinning for supercapacitors. Materials Letters 2012;76:211–4.

[82] Wang CQ, Wang H, Liu YN. Separation of polyethylene terephthalate from municipal waste plastics by froth flotation for recycling industry. Waste Management 2015;35:42–7.

[83] Alam H, Moghaddam G, et al. The influence of process parameters on desulfurization of Mezino coal by HNO_3/HCl leaching. Fuel Processing Technology 2009;90(1):1–7.

[84] Xie G, Song QS, Deng DP, et al. Research progress of needleless electrospinning. Engineering Plastic Application 2014;(6):117–21.

[85] Uslu I, Tunc T, Keskin S, et al. Synthesis and characterization of boron doped alumina stabilized zirconia fibers. Fibers and Polymers 2011;12(3):303–9.

[86] Xing C, Guan J, Li Y, et al. Effect of a room-temperature ionic liquid on the structure and properties of electrospun poly (vinylidene fluoride) nanofibers. ACS Applied Materials and Interfaces 2014;6(6):4447–57.

[87] Ye XY. Electrospinning and functionalization of polypropylene ultrafine fibers. Zhejiang University; 2014.

[88] Pandiyaraj KN, Selvarajan V, Deshmukh RR, et al. Adhesive properties of polypropylene (PP) and polyethylene terephthalate (PET) film surfaces treated by DC glow discharge plasma. Vacuum 2008;83(2):332–9.
[89] Kuwayama M, Vajta G, Kato O, et al. Highly efficient vitrification method for cryopreservation of human oocytes. Reproductive BioMedicine Online 2005;11(3):300–8.
[90] Bhardwaj N, Kundu SC. Electrospinning: a fascinating fiber fabrication technique. Biotechnology Advances 2010;28(3):325–47.
[91] Behnood A, Ghandehari M. Comparison of compressive and splitting tensile strength of high-strength concrete with and without polypropylene fibers heated to high temperatures. Fire Safety Journal 2009;44(8):1015–22.
[92] Lee S, Obendorf SK. Developing protective textile materials as barriers to liquid penetration using melt-electrospinning. Journal of Applied Polymer Science 2006;102(4):3430–7.
[93] Kim J, Hinestroza JP, Jasper W, et al. Effect of solvent exposure on the filtration performance of electrostatically charged polypropylene filter media. Textile Research Journal 2009;79(4):343–50.
[94] Liang C, Hu CY, Yan KL, et al. Influence factors of polypropylene fibers prepared by melts electrospinning. Journal of Textile Research 2016;37(11):14–8.
[95] Zhmayev E, Cho D, Joo YL. Modeling of melt electrospinning for semi-crystalline polymers. Polymer 2010;51(1):274–90.
[96] Liu Y, Deng R, Hao M, et al. Orthogonal design study on factors effecting on fibers diameter of melt electrospinning. Polymer Engineering and Science 2010;50(10):2074–8.
[97] Bognitzki M, Czado W, Frese T, Schaper A, Hellwig M, Steinhart M, et al. Nanostructured Fibers via Electrospinning. Advanced Materials 2001;13(1):70–2.
[98] Hao MF, Liu Y, He XT, et al. Experimental study of melt electrospinning in parallel electrical field. Advanced Materials Research 2011;221:111–6.
[99] Malakhov SN, Belousov SI, Bakirov AV, et al. Electrospinning of non-woven materials from the melt of polyamide-6 with added magnesium, calcium, and zinc stearates. Fibre Chemistry 2015;47(1):14–9.
[100] Hu M, Xu DZ, Guo J. Effects of hyperbranched polymer on rheological and mechanical properties of polypropylene. Journal of Dalian Institute of Light Industry 2007;(02):160–3.
[101] Xiao WQ, Cheng WL, Hu JQ, et al. Research status of hyperbranched polyesters. Material Guide 2008;(04):62–4.
[102] Xin W, Luo YJ, Li XM. Rheological and mechanical properties of stearic acid modified HBPE-PP blends. Polymer Materials Science and Engineering 2008;24(8):66–9.
[103] Cao MG, Yu J, Wu L. Preparation and application of hyperbranched polymer. Plastic 2006;35(3):45–8.
[104] Gao Z, Zhang BY, Liu YH. Synthesis and application of Boltorn hyperbranched polyester. Petrochemical Technology and Application 2004;(06):405–9.
[105] Luo JS, Li P, Yu YH, et al. Characterization of grafted reaction of terminal hydroxyl groups of hyperbranched polyesters by ^{13}C NMR. Plastic 2008;(04):49–52.
[106] Zhao H, Wang L, Luo YJ, et al. Progress in the application of hyperbranched polymers. Engineering Plastics Application 2004;32(5):67–9.
[107] Wang T, Wang YC. Advance on application of hyperbranched polymer. Elastomer 2004;14(4):57–61.
[108] Pohl C, Saini C. New developments in the preparation of anion exchange media based on hyperbranched condensation polymers. Journal of Chromatography A 2008;1213(1):37–44.

[109] XiaLT, LiuY, Ding YM, Yuan HL, et al. Applicationof HyperbranchedPolymers in Melt Electrospinning. Plastics 2012; 41(06):1–3.

[110] Chen ZY, He J, Fengwen Z, et al. Effect of polar additives on melt electrospinning of nonpolar polypropylene. Journal of the Serbian Chemical Society 2014;79(5):587–96.

[111] Ogata N, Lu G, Iwata T, et al. Effects of ethylene content of poly(ethylene-co-vinyl alcohol) on diameter of fibers produced by melt-electrospinning. Journal of Applied Polymer Science 2010;104(2):1368–75.

[112] Dalton PD, Grafahrend D, Klinkhammer K, et al. Electrospinning of polymer melts: phenomenological observations. Polymer 2007;48(23):6823–33.

[113] Tian S, Ogata N, Shimada N, et al. Melt electrospinning from poly(L-lactide) rods coated with poly(ethylene-co-vinyl alcohol). Journal of Applied Polymer Science 2009;113(2):1282–8.

[114] Detta N, Brown TD, Edin FK, et al. Melt electrospinning of polycaprolactone and its blends with poly(ethylene glycol). Polymer International 2010;59(11):1558–62.

[115] Zhmayev E, Cho D, Joo YL. Nanofibers from gas-assisted polymer melt electrospinning. Polymer 2010;51(18):4140–4.

Further reading

[1] Smyth M, Poursorkhabi V, Mohanty AK, et al. Electrospinning highly oriented and crystalline poly(lactic acid) fiber mats. Journal of Materials Science 2014;49(6):2430–41.

[2] Kucinska-Lipka J, Gubanska I, Janik H, et al. Fabrication of polyurethane and polyurethane based composite fibres by the electrospinning technique for soft tissue engineering of cardiovascular system. Materials Science and Engineering 2015;46:166–76.

[3] Liu CL, Guo QG, Shi JL, et al. The curing reaction of phenolic fibers. Chinese Journal of Materials Research 2005;19(1):28–34.

[4] Wang GJ, Wang CM, Li Y. The effect of hyperbranched polyester (BOLTORN^{TM}) on mechanical and rheological properties of rigid PVC. Plastic 2006;(06):15–9.

[5] Xu DZ, Chen YT, Guo J, et al. Rheological property of polypropylene with hyperbranched polyester. Journal of Dalian Polytechnic University 2009;(06):445–7.

Chapter 4

Melt electrospinning in a parallel electric field

Chapter outline

4.1 Introduction

As an efficient, direct, and simple method of producing polymer nanofibers, electrospinning has attracted increasing attention because of its versatility and potential for application in diverse fields [1–3]. Although the setup, materials, and conditions of electrospinning processes vary, fiber instability often occurs, especially when the jet is near the collector. Researchers have different opinions on the reasons for the buckling or "whipping" of the jet. According to Reneker, compressive forces along the axis destabilize buckling when the fluid jet solidifies on a surface [4], as a result of the rheological complexity of the polymer solution and the repulsive forces between the adjacent elements of charges carried by the jet [5]. From Karatay's report whipping is attributed to the chaotic oscillation of a polymer jet [6]. For Hutmacher, instability is

Melt Electrospinning. https://doi.org/10.1016/B978-0-12-816220-0.00004-X

mainly related to the molecular weight, conductivity, and viscosity of a polymer jet [7]. Goki suggested that bending instability depends on polymer molecular weight and concentration [8]. Although all of these authors have verified the accuracy of their inferences via scientific research, we propose that an uneven electric field is another cause of instability. The electric field is a crucial parameter in the electrospinning process, and much research has been conducted on it. For instance, From Erol's [9] report the effect of electric force on multiple jets originates at the needle tip. Yang [10] analyzed the influence of electric field distribution on the electrospinning process and fiber morphology by investigating three types of electric fields (i.e., very nonuniform, slightly nonuniform, and uniform electric fields), and notes that the range of the straight jet path in terms of length is a function of the electric field distribution in a stable electrospinning process. Hao [11] experimentally studied melt electrospinning in a parallel electric field and concluded that many factors affect the stability of the vertical spinning path. We previously examined the effect of polar additives and electric field distribution on the process of melt electrospinning through experiments and simulation, respectively [12,13]. In spite of the different views, the electric field between the small spray nozzle (or tiny needle) and the relatively large collector disk or long roller is apparently nonuniform. In strong electric fields, a polymer polarizes and forms many microdipoles that are irregularly arranged along molecular chains [14]. The misaligned dipoles attempt to maintain their direction by spinning under the influence of the electric field. In uniform fields, the randomly distributed dipoles rotate in different directions; thus, little to no additional or resultant force is generated. In nonuniform fields, the dipoles not only rotate but also move along the field direction. If a fiber component is composed of a large number of dipoles moving in a similar direction, then it is driven to move as well. This movement is enhanced further by the rotation force of the misaligned dipoles in a nonuniform field. Thus, the movements and rotation forces of dipoles strongly contribute to fiber buckling. As a result, a uniform electric field is necessary to reduce fiber instability. To verify this idea, a parallel electric field was generated for real experiments and finite element simulation in this study.

4.2 Method and experiments

4.2.1 Experimental material

Polypropylene (PP6315) was selected for electrospinning because of its good fluidity in molten conditions. The material was obtained from the Shanghai Yishitong material development company. The melt flow rate was 1000 g·$(10\ min)^{-1}$ at (230°C/2.16 kg).

4.2.2 Parallel electrospinning equipment

The equipment for melt electrospinning mainly consists of a spinneret, a heating system, a high-voltage supply device, and a collector [11,12,15–17]. The collector used in this study is an aluminum disk. Another aluminum disk is connected to the spinneret to construct a parallel electric field, as shown in Fig. 4.1B. The diameters of the upper and bottom disks, which are labeled D1 and D2, respectively, may vary. The high-voltage supply device was purchased from the Tianjin High-Voltage Power Supply Plant in China and had a maximum output of +100 kV and a maximum current output of 2 mA. The electrical heating coil that covered the cylinder is custom-built with a power of 300 W. A piston controls the flow rate of the polymer melt.

Fig. 4.1A shows that the spinneret is a steel cone frustum and the polymer melt flows down onto its surface to form many Taylor cones [19,20]. The spinneret hangs from a hole at the center of the upper disk, and the bottom surfaces of the upper disk and the spinneret are on the same horizontal line. The two disks are parallel. The bottom disk is also known as the receiver. The voltages applied to the upper and bottom disks are 0 V and 50 kV, respectively. The custom-made electrospinning system with parallel disks is depicted in Fig. 4.1B.

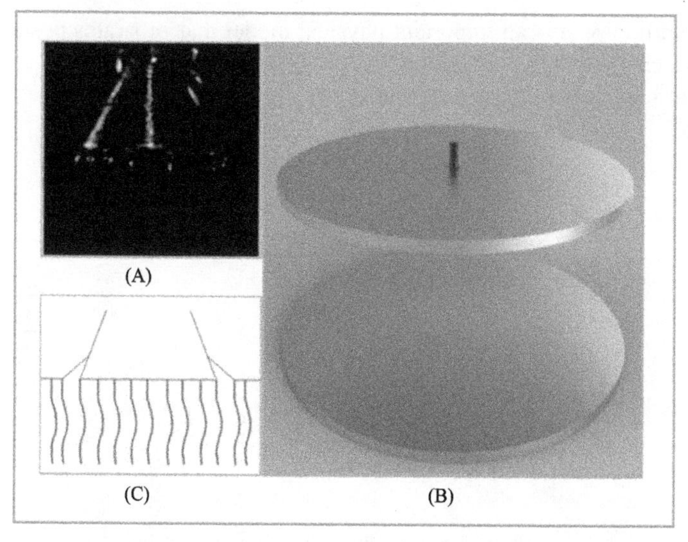

FIGURE 4.1 Setup of electrospinning with parallel disks. Actual spinneret with multiple Taylor cones (A), the parallel disks and the spinneret (B), and mimic diagram of the spinneret with the upper disk (C) [18].

4.2.3 Finite element modeling

Fig. 4.2 displays the finite element model of the parallel-disk system, which contains two 0.5 mm thick parallel disks made of aluminum. $D1$ and $D2$ are 100 and 200 mm, respectively; the distance between the spinneret and the receiver is 120 mm; the diameter of the spinneret bottom measures 10 mm.

Usually, the electric field is mainly located between the spinneret and the receiver in melt electrospinning. In this chapter, the characteristics of the electric field between the two parts are axisymmetric because both parts are also axisymmetric. Thus, the two-dimensional plane finite element model is selected, instead of the three-dimensional solid model, to simplify the simulation. The electric field is closely related to only the spinneret, the bottom disk, and the upper disk connected to the spinneret, as shown in Fig. 4.2. The simulation is based on the following two assumptions to directly and easily obtain results. First, the relative dielectric constant (relative permittivity) [21] of air between the two disks remains stable; the value of the constant is 1 according to other studies [22]. The second assumption is that the electric field is related to only the spinneret and the two disks; therefore, other parts, such as the flying jets, electrical heating coil, temperature-control device, and metal frame of the entire system, do not affect the electric field.

4.2.4 Theoretical analysis

The electric dipole is an important physical model that separates positive and negative charges. When polymer chains are moved into an electric field, the bonds in molecular chains are polarized and form many irregularly arranged

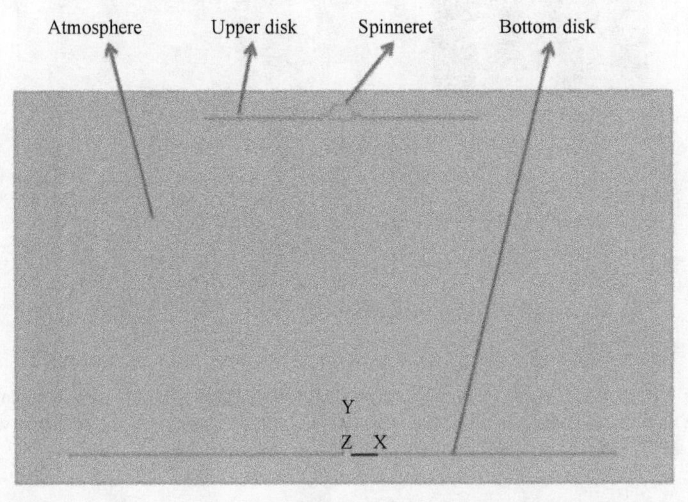

FIGURE 4.2 Finite element model of the setup of electrospinning with parallel disks [18].

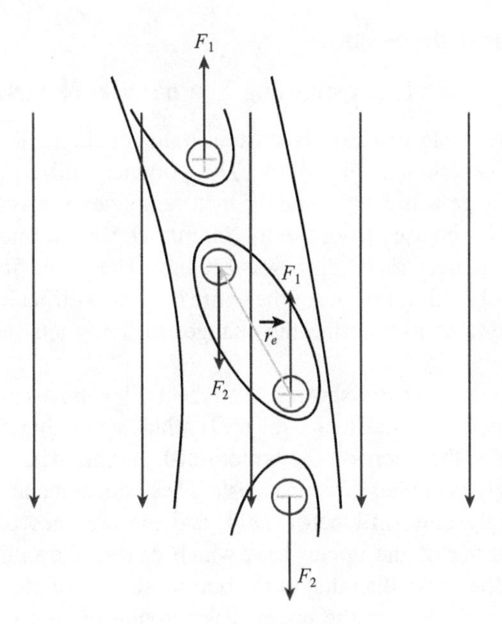

FIGURE 4.3 Model of dipoles during electrospinning in a uniform electric field [18].

microdipoles, as exhibited in Fig. 4.3. To obtain the value of the center dipole, the following formula is used:

$$\vec{P}_e = q \cdot \vec{r}_e \tag{4.1}$$

$$\vec{F} = \nabla(\vec{P}_e \cdot \vec{E}) \tag{4.2}$$

$$\vec{M} = \vec{P}_e \times \vec{E} \tag{4.3}$$

where, q is the quantity of electric charge; $\vec{r}_e$ is the radius vector from negative to positive charge; $\vec{P}_e$ is the dipole moment; $\vec{E}$ is the strength of the electric field; $\vec{F}$ is the resultant force; and $\vec{M}$ is the torque. In this analysis, the effect of the other dipole on the center dipole is neglected. If the dipole in a uniform electric field is not parallel with the electric field line, then $\vec{E}$ is constant and $\vec{F}$ is zero. However, $\vec{M}$ is not zero, thus causing the dipole to rotate. Nearby dipoles have either a similar or inverse central rotation trend, and little to no additional rotational force is exerted. In a nonuniform electric field, $\vec{E}$ varies with different spatial positions. Both $\vec{F}$ and $\vec{M}$ are not zero, therefore, the dipole moves in a translational and rotational manner. Neighboring dipoles have similar motion trends, and the motion of the jet segment is stimulated, thus, jet instability is induced.

4.3 Results and discussion

4.3.1 Experimental electrospinning in a parallel electric field

A parallel electric field platform is custom-built for the melt electrospinning experiment, as depicted in Fig. 4.4A. The polymer utilized polypropylene; the spinning voltage was 50 kV, and the melt temperature was 230°C. *D*2 was 200 mm, and *D*1 increased from 50 to 200 mm at 50-mm intervals. The distance between the two parallel disks was 120 mm. The jet or fiber produced is very fine, thus, obtaining images of the entire trace was difficult. Therefore, we drew mimic diagrams to describe the changes in flying jets between the two disks, as shown in Fig. 4.4B.

The experimental results showed that the falling traces of the jets as a whole straightened with the increase in *D*1, thus stabilizing the falling jets. The falling area of the fibers on the bottom disk shrank, whereas the polymer melts increasingly occupied the upper disk. These phenomena were attributed to the increasingly uniform electric field, and the presence of several small girders in the center of the upper disk, which caused the falling area of the polymer melt to be larger than that of the bottom surface of the spinneret. Parts of the polymer melt flow to the upper disk through these girders, which are connected to the surface of the spinneret, as displayed in Figs. 4.1C and 4.4A.

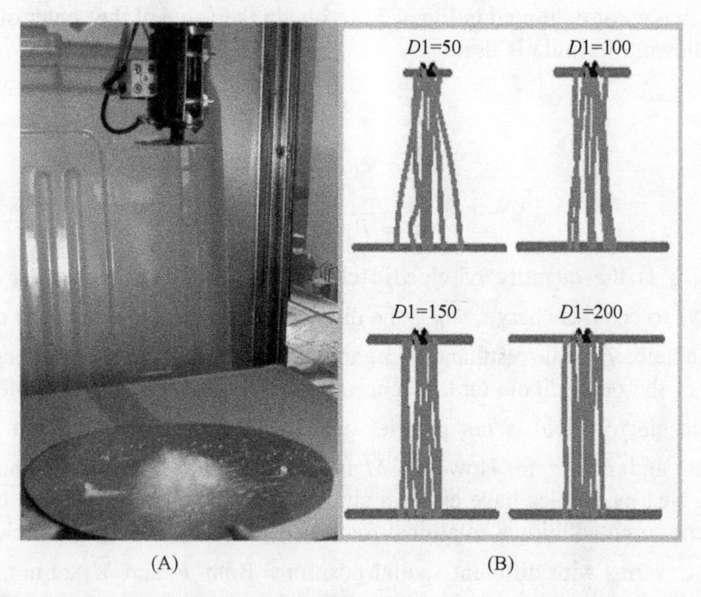

FIGURE 4.4 Setup of melt electrospinning with parallel disks (A) and illustrations of changes in traces with different upper disk diameters (B) [18].

Other undiscovered factors may also contribute to these findings. Therefore, finite element simulation was conducted to study the process of melt electrospinning in a parallel electric field and enrich our understanding of the occurrences observed above.

4.3.2 Finite element simulation of the electrospinning process in a parallel electric field

4.3.2.1 Changes in the parallel electric field with the diameter of the upper disk

In the simulation model, the diameter of the bottom disk was fixed whereas that of the upper disk varied. Fig. 4.5 exhibits the cloud chart of electric potential in different situations and illustrates the changes in the electric field with the increase in the diameter of the upper disk. When $D1 = 0$ or when an upper disk is absent, the electric field between the spinneret and the bottom disk is completely uneven. However, the electric field changes when $D1$ increases from 50 mm to the final value of 200 mm, which is equal to $D2$. The color belts that represent the contour of the electric field flatten in the center, especially when both disks have the same diameter.

Electrospinning is typically conducted in nonuniform electric fields. The polarized or charged jet easily drifts from the previously straight jet path when random force is generated. This force does not accord with the centerline between the spinneret and the center of the bottom disk and originates from many factors, such as the disturbance of the surrounding air, the flow of wind in the electric field, and unbalanced movements of uneven charges on the melt jet. When the charged jet is in motion in an even electric field, as given by the furthest right image in Fig. 4.5, the disturbance is largely reduced because the evenness of the field prevents the generation of random forces. The wind in the electric field is also weaker than that in the spinneret-disk electric field, and the electric force drives the jet directly to the bottom.

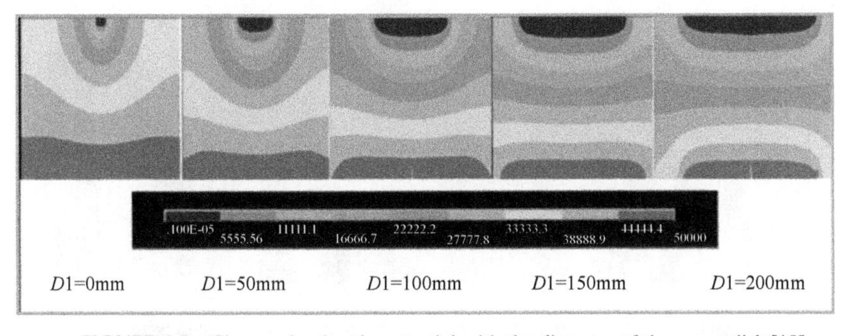

FIGURE 4.5 Changes in electric potential with the diameter of the upper disk [18].

4.3.2.2 Changes in the parallel electric field with the distance between two parallel disks

The distance between $D1$ and $D2$ changes when both parallel disks measure 50 mm and the voltage applied to them was 50 kV. The changes in the electric field with distances of 50 mm, 80 mm, 100 mm, 120 mm, and 200 mm were analyzed, as shown in Fig. 4.6. The shift in color from blue to red denoted the enhancement of the electric field intensity from weak to strong. The images in Fig. 4.6 were arranged exactly according to height for easy comparison and to illustrate the decrease in electric field intensity in the jet trace areas with the increase in the distance between the two parallel disks. The increasing distance reduced the connecting region of the cyan area, and this area disappered when distance reaches 120 mm. In this figure, the red arrows indicate the strong electric field intensity at the start point. The electric field intensities at the two ends of the disk are the strongest compared with those of other areas. This strong intensity is attributed to the concentration of the electric field on the sharp or end part of the metal conductor. To obtain detailed data, the electric field intensity in the path of the black dotted line in Fig. 4.6E was determined from the simulation results and displayed in Fig. 4.7. This path begins from the center of the spinneret bottom and leads to the collector center. All of the curves resemble saddles, thus indicating that electric field intensity rises sharply near the collector. The intensity then decreases slowly until it reaches the center of the path and then increases again because of its proximity to the upper disk. Finally, the electric field intensity greatly decreases near the spinneret bottom. The sudden initial and final changes may be attributed to the increased concentration of intensity on either the spinneret bottom or the disk edge. The intensity on the disk edge is stronger than that on the center of the disk or on the spinneret bottom. With respect to the middle part of the path, the data are in accordance with the chart exhibited in Fig. 4.6, thus showing that the intensity near the disk is high, and vice versa.

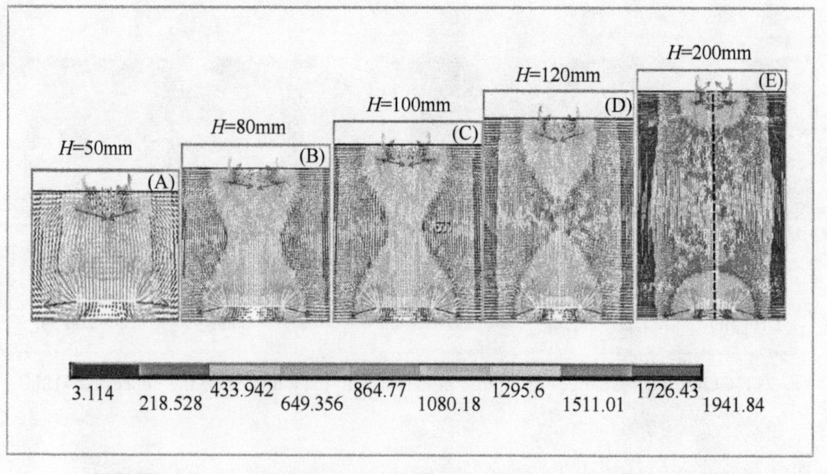

FIGURE 4.6 Vector diagram of intensity distribution in the electric field [18].

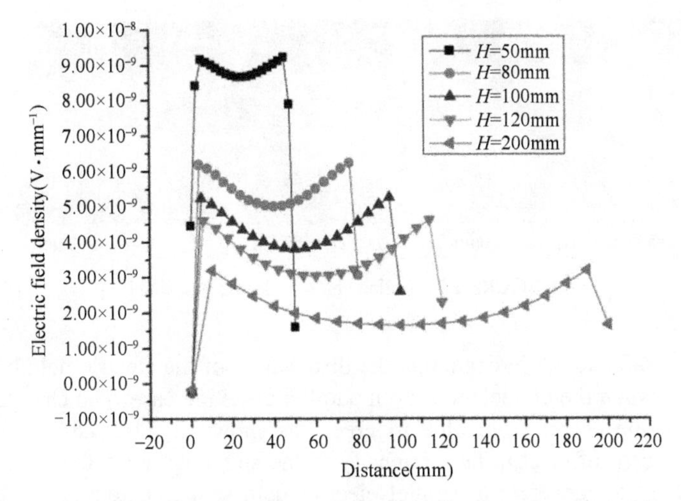

FIGURE 4.7 Electric field intensity in the vertical path leading from the center of the spinneret bottom to the collector center [18].

4.3.2.3 Changes in the electric field with the diameters of the two parallel disks

The changes in the electric field with diameters of 50, 100, 150, and 200 mm were analyzed separately using the finite element method when the diameters of the two parallel disks were similar and the distance between them was 120 mm. Fig. 4.8 depicts a vector diagram of the distribution of electric field intensity. As observed in this figure, electricity is always concentrated at the ends or edges of the disks regardless of disk diameter. With the increase in diameter, the distance between the ends of the disk and the spinneret widens and the end effect of electricity on the fibers weakens, thus causing the electric field to even out. Fig. 4.9 shows the isoline chart of electric potential.

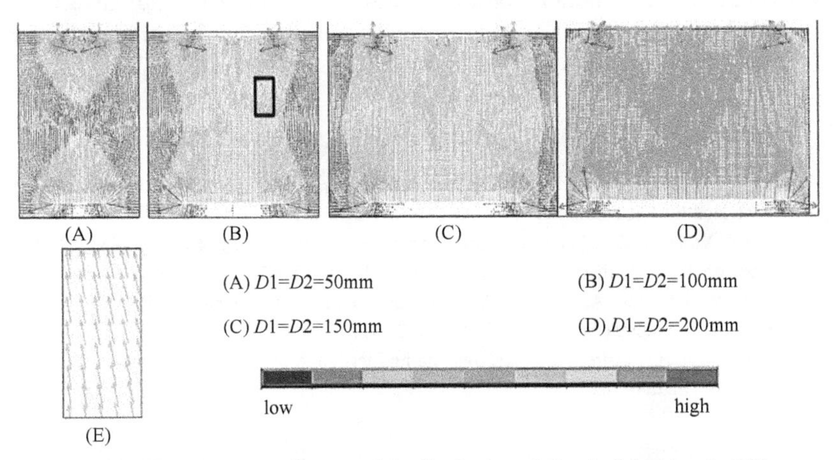

FIGURE 4.8 Vector diagram of the distribution of electric field intensity [18].

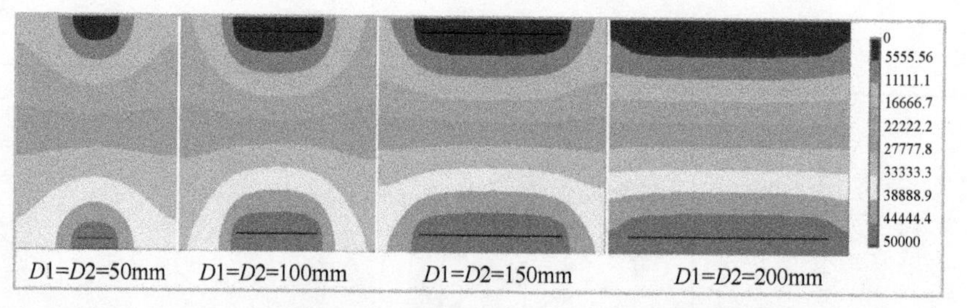

FIGURE 4.9 Isoline chart of electric potential [18].

Figs. 4.8 and 4.9 indicate that the distribution of the electric field becomes uniform when the diameters of both parallel disks increase. The change in the color of the center from blue to green, as shown in Fig. 4.8, denotes the enhancement of electric field intensity in the spinning area. However, not all parallel plates generate a parallel electric field at the center. The uniformly parallel electric field is formed only when the distance between the two disks is smaller than the diameters of both disks.

4.4 Conclusion

This chapter verifies that a uniform electric field can reduce jet whipping in melt electrospinning through experiments and finite element simulation on a parallel electric field and that dipoles greatly contribute to fiber buckling. When the diameters are fixed, the electric field intensity in the jet trace areas decreases if the distance between the two parallel disks increases. The electric field intensity on the sharp or end part of the metal disk is also always stronger than that at the center of the disk or at the spinneret bottom because intensity concentrates easily at the edge of the disk. The electric field becomes even when the area of the bottom disk is fixed and the diameter of the upper disk increases. When the diameters of both parallel disks are similar, the distribution of the electric field becomes uniform, and the electric field intensity in the spinning area increases as the diameter increases. Thus, the parallel electric field effectively reduces jet buckling in melt electrospinning.

References

[1] Du Q, Harding DR, Yang H. Helical peanut-shaped poly(vinyl pyrrolidone) ribbons generated by electrospinning. Polymer 2013;54(25):6752–9.

[2] Liu Z, Xiong L, Yin Y, et al. Control of structure and morphology of highly aligned PLLA ultrafine fibers via linear-jet electrospinning. Polymer 2013;54(21):6045–51.

[3] Agarwal S, Greiner A, Wendorff JH. Functional materials by electrospinning of polymers. Progress in Polymer Science 2013;38(6):963–91.

[4] Reneker DH, Yarin AL. Electrospinning jets and polymer nanofibers. Polymer 2008;49(10):2387–425.

[5] Reneker DH, Yarin AL, Fong H, et al. Bending instability of electrically charged liquid jets of polymer solutions in electrospinning. Journal of Applied Physics 2000;87(9):4531–47.
[6] Karatay O, Dogan M. Modelling of electrospinning process at various electric fields. Micro and Nano Letters IET 2011;6(10):858–62.
[7] Hutmacher DW, Dalton PD. Melt electrospinning. Chemistry - An Asian Journal 2011;6:44–56.
[8] Eda G, Liu J, Shivkumar S. Flight path of electrospun polystyrene solutions: effects of molecular weight and concentration. Materials Letters 2007;61(7):1451–5.
[9] Jentzsch E, Ömer G, Öznergiz E. A comprehensive electric field analysis of a multifunctional electrospinning platform. Journal of Electrostatics 2013;71(3):294–8.
[10] Yang Y, Jia Z, Liu J, et al. Effect of electric field distribution uniformity on electrospinning. Journal of Applied Physics 2008;103(10):89.
[11] Hao MF, Liu Y, He XT, et al. Experimental study of melt electrospinning in parallel electrical field. Advanced Materials Research 2011;221:111–6.
[12] Wang X, Liu Y, Zhang C, et al. Simulation on electrical field distribution and fiber falls in melt electrospinning. Journal of Nanoscience and Nanotechnology 2013;13(7):4680–5.
[13] Chen ZY, He J, Fengwen Z, et al. Effect of polar additives on melt electrospinning of nonpolar polypropylene. Journal of the Serbian Chemical Society 2014;79(5):587–96.
[14] Service RF. Nanolasers. Smallest of the small. Science 2010;328(5980):811.
[15] Cao K, Liu Y, Olkhov AA, et al. PLLA-PHB fiber membranes obtained by solvent-free electrospinning for short-time drug delivery. Drug Delivery and Translational Research 2018;8(1):291–302.
[16] Viswanadam G, Chase GG. Modified electric fields to control the direction of electrospinning jets. Polymer 2013;54(4):1397–404.
[17] Eichholz KF, Hoey DA. Mediating human stem cell behavior via defined fibrous architectures by melt electrospinning writing. Acta Biomaterialia 2018;75:140–51.
[18] Liu Y, Li X, Ramakrishna S. Melt electrospinning in a parallel electricfield. Journal of Polymer Science Part B: Polymer Physics 2014;52(14):946–52.
[19] Yarin AL, Koombhongse S, Reneker DH. Taylor cone and jetting from liquid droplets in electrospinning of nanofibers. Journal of Applied Physics 2001;90(9):4836–46.
[20] Moroshkin P, Leiderer P, Möller TB, et al. Taylor cone and electrospraying at a free surface of superfluid helium charged from below. Physical Review E 2017;95(5–1):053110.
[21] Luo CJ, Stride E, Edirisinghe M. Mapping the influence of solubility and dielectric constant on electrospinning polycaprolactone solutions. Macromolecules 2012;45(11):4669–80.
[22] Duan HW, Jiang JG. Finite element analysis and optimization of electric field structure of new-type electrospinning machine. Advanced Materials Research 2011;279:214–8.

Chapter 5

Dissipative particle dynamics simulation on melt electrospinning

Chapter outline

5.1 Introduction

Dissipative particle dynamics (DPD) is a mesoscopic simulation algorithm used to simulate complex fluid behavior. It is used to solve the fluid problem

Melt Electrospinning. https://doi.org/10.1016/B978-0-12-816220-0.00005-1

on the mesoscopic time and space scale that cannot be solved by the lattice automaton method and the molecular dynamics.

Both the dissipative particle dynamics method and the molecular dynamics method are based on the dynamics simulation method of Newton's equation of motion. The DPD method, to simulate the movement of particles in the system, follows Newton's law [1–5]:

$$\begin{cases} \dfrac{\partial r_i}{\partial t} = v_i \\ m_i \dfrac{\partial v_i}{\partial t} = f_i \end{cases} \tag{5.1}$$

The force of particle i is f_i:

$$f_i = \sum_{j \neq i} \left(F_{ij}^C + F_{ij}^D + F_{ij}^R \right) \tag{5.2}$$

The dissipation force F^D between the particles in the DPD system is used to describe the viscous resistance between the particles in the system. The random force F^R is used to describe the effect of thermal motion of the particles in the system.

However, the dissipative particle dynamics method differs from the molecular dynamics method in that the particles represent, not real atoms, but can be thought of as fluid particles composed of some molecules. If a polymer chain is considered, the DPD particle is composed of several monomer units or even a few chains [6–9].

DPD is a mesoscopic simulation method developed on the basis of the molecular dynamics (MD) method and the LGA (Lattice-Gas Automata) model. Combining the advantages of the MD method and the LGA method, the lattice problem in the LGA is avoided while retaining its simplicity. At the same time, the calculation step size of DPD is much larger than that of MD [10]. The basic feature of DPD simulation is to divide the fluid into small quantities. The mass and momentum in the system are conserved, but the energy is not conserved.

The fiber diameter of the melt electrospinning is generally in the order of micrometers, and a few can reach the nanometer scale, which is physically a mesoscopic scale. Therefore, using the mesoscopic simulation method to study the melt electrospinning drop process, fiber change, molecular chain orientation, etc., can more accurately reflect the nature of the spinning phenomenon. It also reveals the unwinding and orientation changes of polymer chains, fiber formation, and motion laws under multifield coupling.

In order to solve the serious phenomenon of the entanglement of the melt electrospinning chain and to understand the physical movement of the molecular chain during the fiber jet process, our laboratory further applied and developed the description of melt electrospinning by DPD. The dynamic movement process of molecular chains in melt spinning and its influence on

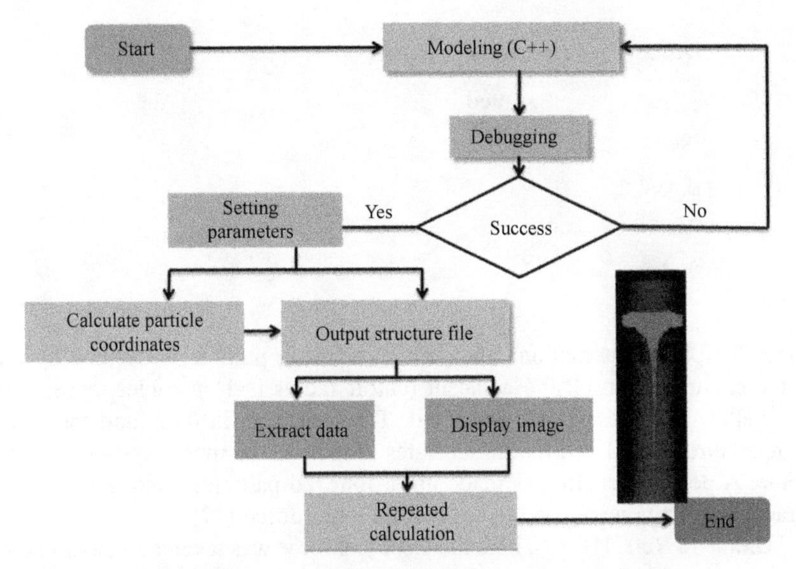

FIGURE 5.1 A flow diagram of the changed electric field [11].

macroscopic morphology were studied. The application process is as follows. First, the variables are integrated into the calculation process of the entire jet motion through the C++ calculation. Then, according to the output structure file, the coordinates of each polymer particle are extracted. Finally, the molecular chain situation and the data of jet diameter and falling velocity are studied. Fig. 5.1 is a flow chart for the application of a variable electric field to the melt electrospinning process.

In this study, the model system is the same as those used in previous studies [12,13]. The simulation box, subjected to the usual periodic boundary conditions, contains 24,000 particles in $10 \times 10 \times 40$ cubes. There are three kinds of particle in the box; those found in the environment or air (blue particles), the polymer melt (red particles), and the spinning platform (yellow particles). The air is composed of 19,200 blue particles, the melt consists of 3353 red particles, and the spinning platform contains 1447 yellow particles. The yellow particles are fixed and do not move along with the other particles. The red particles are connected by several springs, whereas the other particles are single, similar to the model systems described in Refs. [14,15]. The three kinds of particles have the same density $\rho = 0$ (each unit cube contains six particles). The parameters of conservative forces are chosen in accordance with the data in Table 5.1. The parameter assigned between yellow particles and the others is 80, to prevent moving particles from entering or threading into the solid spinning platform. There is no difference between the three particles in

TABLE 5.1 Parameters of conservative forces particles[16].

	Red	Blue	Yellow
Red	40	30	80
Blue		12.5	80
Yellow			12.5

their dissipative interaction. The velocity of each particle is decided by the force endured from all the adjacent (cutoff radius is 1) particles, except for the yellow particles, which are fixed. The time step is 0.02, and the basic temperature is 0.10. All the initial states of melt in the model systems are the same. A polymer chain was made up of four red particles, among which the Fraenkel springs were used to express the bond force [17].

Unlike in Refs. [18,19], here the electrical force was taken as an extra body force rather an interaction between target polymer chains or different particles. The electric field existed only between the bottoms of the spinning platform and the lowest boundary of the simulation system, in which the boundary was taken as the fiber collector, as shown in Fig. 5.2. Here, according to experimental observation and Ref. [20], the electric field is thought to be not even, as

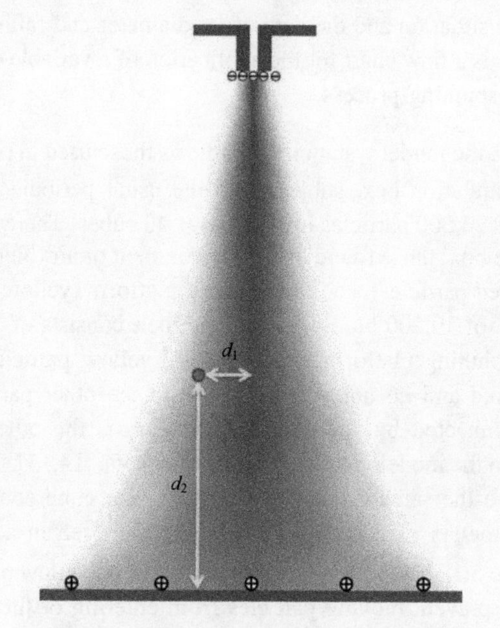

FIGURE 5.2 Schematic of an electrospinning field [16].

shown with blue color. Herein, dark blue means a strong electric field, and vice versa. Accordingly, the tentative electrostatic force is given by:

$$F_i = -c(A - d_1 + d_2)\boldsymbol{e} \tag{5.3}$$

where, c is a coefficient and can be used to adjust the force at will; A is a constant used to decide the basic amount of electrostatic force (15 was found to be a suitable value for the present simulation system); d_1 denotes the distance of the particle to the central axis of the simulation system, a small value means a big electrostatic force; d_2 represents the distance to the bottom of the spinning platform, a big value means a big electrostatic force; $\boldsymbol{e}$ is a unit vector and means that the electrostatic force always points to the bottom. Usually, gravitation is weak because the spinning fiber is very thin, and the electrostatic force is the main driving force. Therefore, gravity is not considered in the present model system.

5.2 Differential scanning calorimetry simulation under different electric fields

5.2.1 Electrostatic field

5.2.1.1 Simulation of dropping trace in melt electrospinning

The dropping trace is an important character of electrospinning. Researchers have found that the dropping trace is more unstable in solution electrospinning, especially in the field near the receiver plate, as shown in Fig. 5.3C. The reason for the instability possibly contributes to the small diameter of the fiber, which is easily affected by weak disturbance of air or the

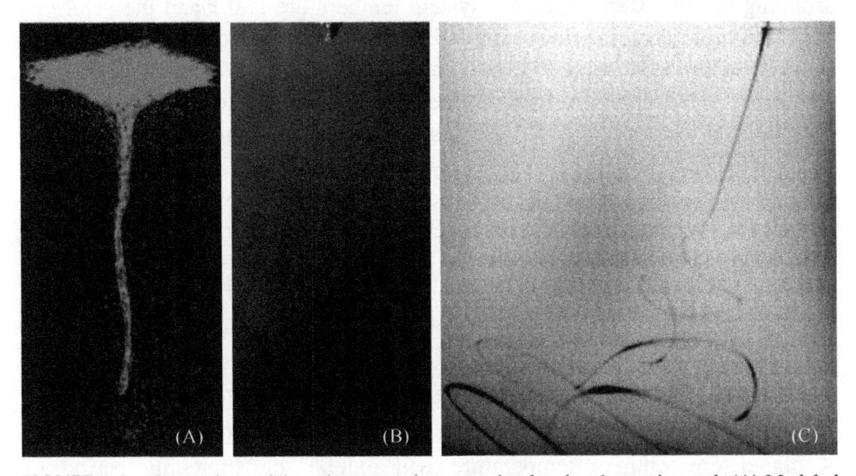

FIGURE 5.3 Comparison of dropping traces between simulated and experimental. (A) Modeled trace, (B) snapshot of melt electrospinning trace, and (C) snapshot of solution electrospinning trace [20].

electrostatic field. In contrast, the trace is more stable to melt electrospinning. It can be observed directly when the fiber diameter is large, as shown in Fig. 5.3B. Through adjusting the interactive parameter of the simulation system, the dropping trace was modeled and compared with the experimental results to justify that the new simulation system is suitable for the melt electrospinning process. As shown in Fig. 5.3A, the modeled dropping trace is similar to that in the melt electrospinning experiment (Fig. 5.3B) and the prehalf trace of the solution electrospinning experiment (Fig. 5.3C).

5.2.1.2 The relationship between electrostatic force and dropping velocity of the fiber

Under a high electric field, both polar and nonpolar polymers are polarized and produce electric charges in their inner and surface components (electric dipoles [21]). Charged polymer melt under electrostatic force is rapidly stretched and then falls to the receiving electrode or plate. During the drop process, the polymer melt is solidified because of heat loss. Theoretically, the higher the electrostatic force, the greater the polarization that the polymer will undergo because the higher charge on the melt means a higher electrostatic force was applied to it. The process of melt electrospinning and the value of electrostatic force are closely related. The relationship between voltage and fiber diameter has been investigated through experiments [22,23], but it is difficult to measure the relationship between the electrostatic force and the fiber dropping rate. Therefore, a suitable simulation system was established to study the effect of electrostatic force on dropping velocity. Without changing other conditions, the electrostatic force coefficient can be modified to obtain different values of electrostatic force. Fig. 5.4 shows the morphology of fiber dropping in 3000 steps when the system temperature is 0.1 and the values of electrostatic force coefficient c are 0.65, 0.7, 0.75, 0.8, and 0.85, respectively.

As can be seen from Fig. 5.4, increasing electrostatic force coefficient accelerates the dropping velocity of the fiber. The average dropping velocity is calculated as shown in Fig. 5.5. Each set of data in this figure is an average of 10 repeated calculations because the dynamic process includes random factors. Fig. 5.5 shows that when the coefficient of the electrostatic force (which represents the electrostatic force directly) increases, the dropping velocity of the fiber gradually rises. However, the change in velocity is not linear with the increase in electrostatic force. When the coefficients of electrostatic force reach a certain value, the dropping velocity rapidly increases.

5.2.1.3 Variations in temperature lead to changes in the fiber structure

A higher environment temperature will extend the stretching period of fiber melt. Therefore, it is necessary to obtain thinner fibers in the electrospinning experiment. However, the heat transfer between melt and air is very

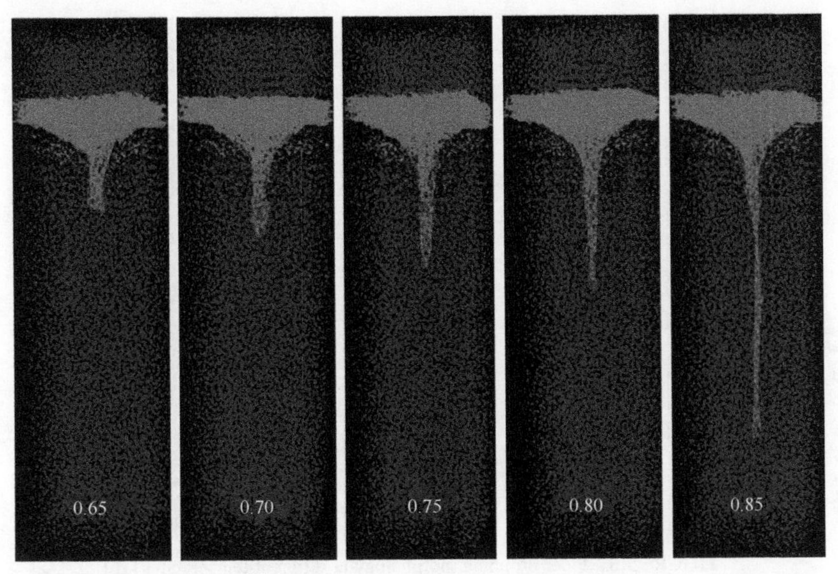

FIGURE 5.4 Relationship between electrostatic force coefficient and dropping velocity [16].

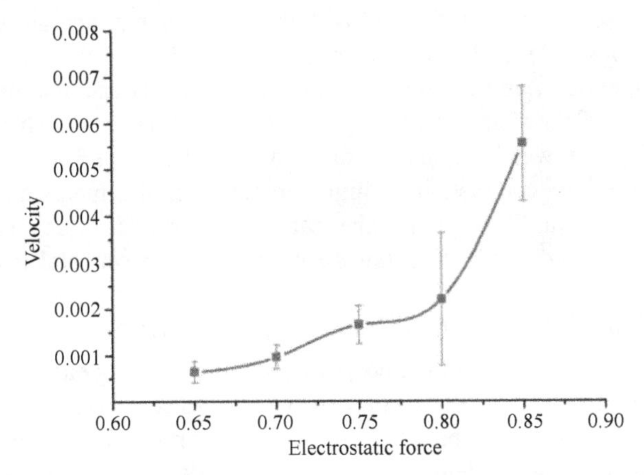

FIGURE 5.5 Changes of dropping velocity with coefficient of electrostatic force [16].

complicated [24,25], and it is difficult to simulate accurately. Also, the change of melt temperature is the key to affecting the morphology of the last fibers. This chapter has therefore adopted the assumption that there is no heat transfer to avoid the above problem and focused only on the melt temperature varying. The fiber dropping processes were simulated when system and fiber melt temperatures were 0.09, 0.10, 0.11, 0.12, and 0.13 using an electrostatic force coefficient of 0.75, and with other conditions unchanged. The results are shown in Fig. 5.6.

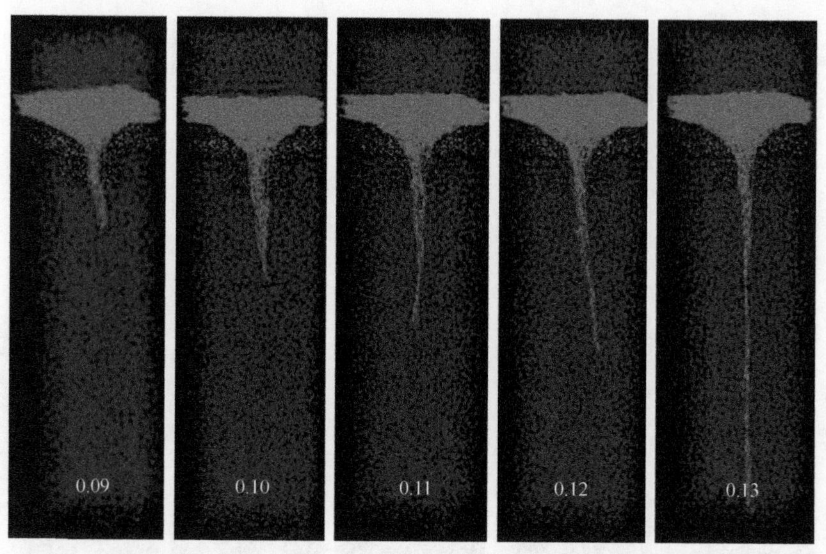

FIGURE 5.6 Fiber dropping morphology at different temperatures [16].

It can be seen from Fig. 5.6 that with the increase in temperature, the fiber gradually drops faster and its diameter becomes smaller. Particles move slowly in low temperature, and a number of molecular chains accumulate together, yielding larger fiber diameters. In experiments, when the temperature was low, the melt viscosity was high, and fluidity was low. The melt flow formed thick fibers quickly. By contrast, in a high temperature, the melt can be easily stretched into thin fibers under the same electrostatic force. The above-mentioned modeled results are consistent with the experimental results [22].

5.2.1.4 Chain length during the dropping period

During melt electrospinning, the polymer melt under strong electrostatic force was stretched and its descent to the receiver plate was accelerated, thus, polymer chains are drawn continuously. Experiments cannot provide variations of molecular chains, but the lengths of molecular chains can be simulated. When the electrostatic force coefficient was 0.75 and the temperature was 0.12, three molecular chains were selected. The end-to-end distances or chain lengths of the three chains were calculated. The change in the distances during the dropping process is shown in Fig. 5.7.

As shown in Fig. 5.7, chain lengths became longer with increasing steps, but this was not uniform throughout the entire process. Initially, the chain lengths became shorter, mainly because the chains were initially connected in the sequence of the particle ranking number, not the adjacent particles in the simulation system. Most chains will be very long because the particles were randomly dispersed in the melt area of the system at the beginning. When the

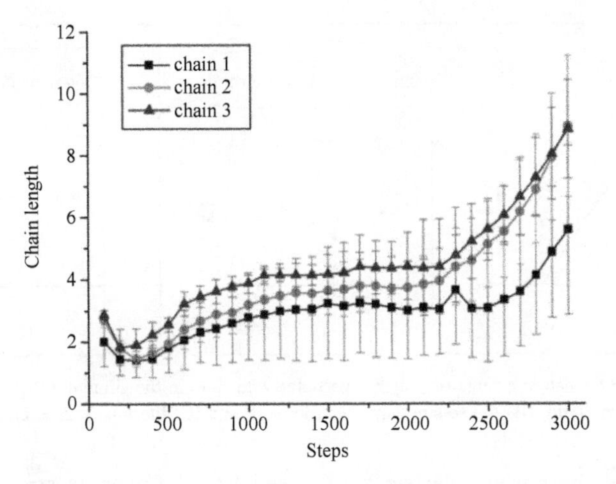

FIGURE 5.7 Changes of chain lengths of three dropping molecules [16].

system began to evolve, the excessively lengthy chains contracted or shortened. The long middle part shows that the increase in chain length is gradual. On the one hand, this can be explained by the fiber dropping velocity increasing continuously and molecular chains becoming longer at the same time. On the other hand, high dropping velocity leads to high air resistance, resulting in accelerated speed deceleration. Therefore, chain lengths increase slowly. The last part of the figure shows that the chain lengths rapidly increase, which is in accord with experiment phenomena [26] and also complies with Fig. 5.5. The fast increase in chain length originates from the constantly increasing electrostatic force. The polymer is stretched extensively and chain lengths increase quickly.

5.2.2 Pulsed electric field

In normal direct current (DC) electric fields, the force that one polymer particle undergoes, from the nozzle to the receiving plate, is shown in Fig. 5.8A. The force decreases with an increase of the flying distance because the strength of the electric field declines quickly [16]. After including the control factor, a square wave, as shown in Fig. 5.8B, the electric field force changes over time, as shown in Fig. 5.8C. The electric field force is no longer continuous. At one moment, the particles were subjected to an electric field force as in a normal DC electric field. The length of this moment is equal to the pulse width. The next moment, the particles are not subject to any force. These two moments make up one period, and this pattern repeats. We define this as the pulsed electric field. In this chapter, we focus on how the pulse width, period, and duty cycle of the pulsed electric field affect the melt electrospinning process compared with the normal DC electric field. Here, the

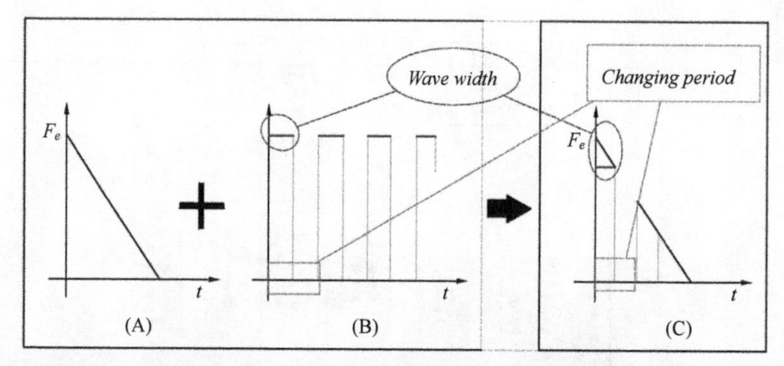

FIGURE 5.8 Changing force of polymer particles with time in the simulation system. (A) The normal electric field, (B) the square wave electric field, and (C) the final pulsed electric field [11].

strength of the normal electric field is equal to the average strength of the pulsed electric field.

The polymer jet in the pulsed electric field had the same kind of charge, which means the direction of the electric field force is constant, but the electric field strength changes over time. The pulsed electric field is somewhat similar to the AC electric field in which the net charge of fiber jets is 0, but their impact on the jetting fiber is completely different [27,28]. The biggest difference of the pulsed electric field from the AC electric field is that the jet is not electrically neutral. The molecular chains in the pulsed electric field have a like charge, so adjacent chains repel each other. Their relationship is shown in Fig. 5.9. Thus, the molecular chains unwrap more easily in the pulsed electric field than in the AC electric field, resulting in a lower viscosity of the polymer melt. Thus, we predict that the pulsed electric field is capable of producing fibers with a small diameter.

In the melt electrospinning simulation system (Fig. 5.10), the jet is generated with a certain initial velocity and then stretched with the electric

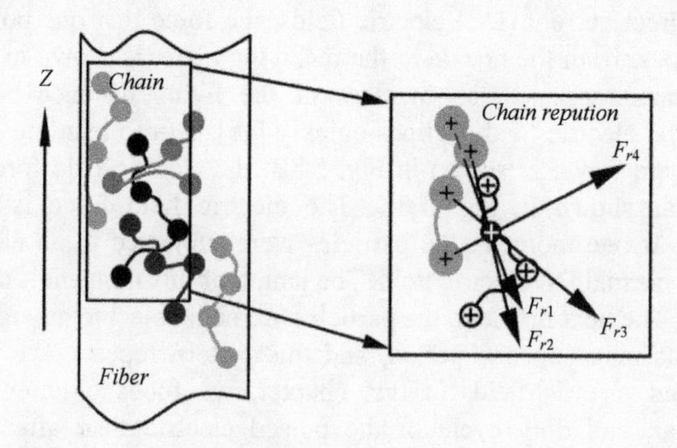

FIGURE 5.9 Charge repulsion between adjacent molecular chains.

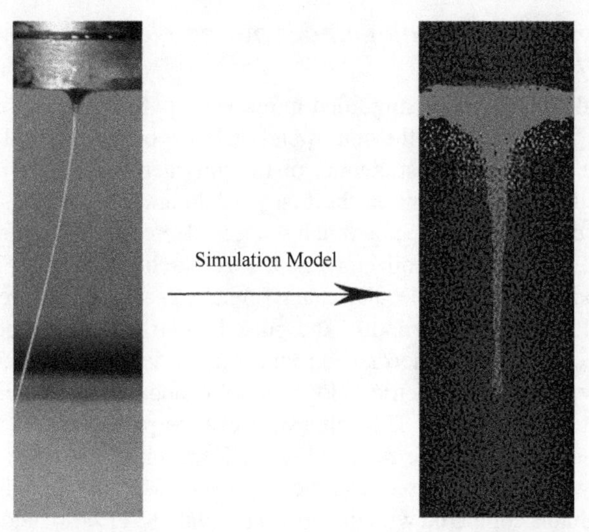

FIGURE 5.10 Experimental (left) and simulated (right) jets in melt electrospinning [11].

field force (Fig. 5.10). Polymers in the barrel are unaffected, only those flowing out of the nozzle experience the electric field force. The values of the parameters in the simulation system do not represent the experimental values directly but reflect the same trend as the actual experiment.

In this chapter, a pulsed electric field was applied between the spray nozzle and receiver in the melt electrospinning process. The parameters governing melt electrospinning differ from those of solution electrospinning. We focused on the effects of the duty cycle and pulse width of the pulsed electric field on fiber diameter and molecular stretching.

The polymer particles experienced an intermittent force, which means that the continuous melt electrospinning process was discretized into many intermittent steps in the simulation. The greater the numbers of intermittent steps, the shorter the cycle is, which means that the electric field changes more quickly.

The internal structure of the melt electrospinning fibers can be determined through some characterizations, such as crystal structure probe by differential scanning calorimetry (DSC) or X-ray diffraction (XRD), the molecular chain orientation test, etc. [29,30]. The data were obtained from the fibers that we eventually collected. Before the fiber is tested, some changes occurred, which make the fiber different from the jet just out of the nozzle. In addition, it is difficult to perform real-time measurement during the spinning process. Molecular chain stretching is an intuitive physical quantity that can reflect the internal structure of a jet. The DPD method provides knowledge on the molecular chain stretching of a fresh jet or allows measurement of the stretch at any period during the process.

5.2.2.1 Molecular stretching under pulsed electric field with different cycles

The molecular chains were simplified into several particles strung together by some bond. Here, we used the end-to-end distance of the molecular chain to characterize the molecular stretching of the polymer chains. The end-to-end distance is the distance between the first particle and the last particle of one molecular chain. Three molecular chains were selected randomly from the jet, just before reaching the collector, as the research objectives. They were recorded and labeled as molecular chain 1, chain 2, and chain 3, respectively. In Fig. 5.11, the change period of the pulsed electric field is plotted on the transverse coordinate. The normal constant electric field is expressed as period 0, which means that the electric field does not change while the jet flies from the nozzle to the collector. The value of a change period means the change cycles of the pulsed electric field in a certain spinning time. The simulation results shows that the pulsed electric field increase the stretching of the molecular chains compared with that in the normal DC electric field. However, the effect of the period change on the molecular stretching is not obvious. With the increase of the period number, the amount of molecular stretching decrease and increase, then repeats, showing a fluctuating change. The amount of change is different; the maximum molecular stretching occurs at period 8. The above phenomena are related to the molecular force. When in the normal DC electric field, the force that a given particle experiences is continuous, and the force difference, between the segments of a molecular chain, is small. In the pulsed electric field, the force that a given particle experiences is nonconstant. At one moment, the particle is accelerating because the electric field force is larger. The next moment, the electric field force is 0, and the particle is decelerating due to resistance.

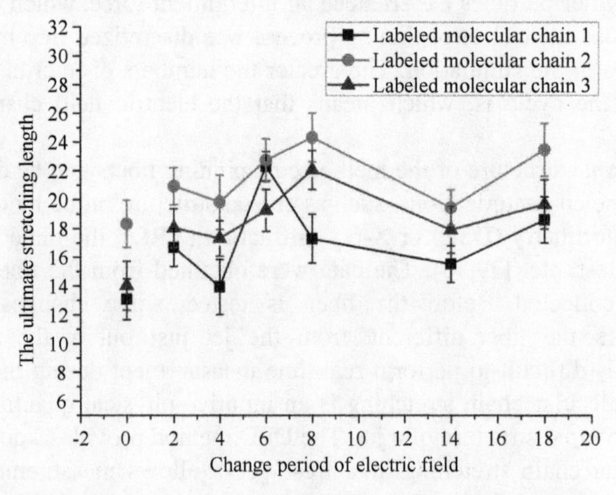

FIGURE 5.11 Changes to the ultimate stretching length of chains with cycles of the electric field [11].

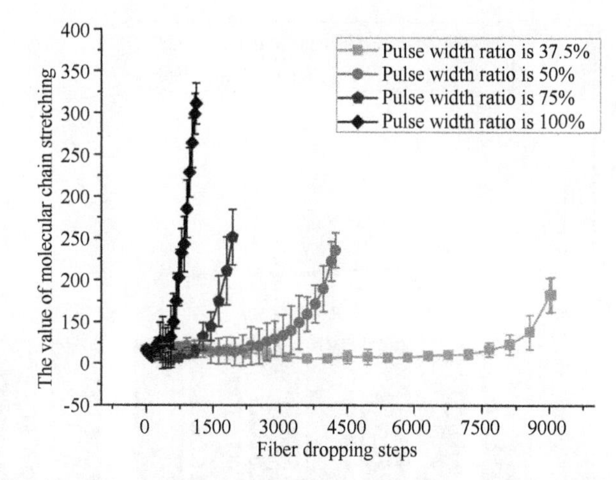

FIGURE 5.12 Fiber stretching under different duty cycles over the same distance [11].

The movement difference between the segments is increased, and the molecular stretching is greater.

5.2.2.2 Effect of pulse width on molecular stretching

When selecting the period of a pulsed electric field, a longer pulse width implies a longer electric field force time and a shorter intermittent time. The electric field force affected the arrangement of molecular chains. As shown in Fig. 5.12, the stretching of a molecular chain was examined through the entire jetting process with period 8, and the duty cycle ranging from 37.5% to 100% (Fig. 5.12). Obviously, the pulse width affects the falling time of a jet in the same distance. The falling time decreases with the increasing pulse width, and molecular stretching becomes slightly larger with the increasing pulse width, but the stretching is far less than the amount of stretching in the full pulse width. This is because a longer pulse width provides greater continuous action time of the electric field force, which provides a longer stretching time.

5.2.2.3 Jet diameter in a pulsed electric field

In our simulation system, the molecular chain is relatively short and only one polymer jet erupted from the spinneret [31]. The fiber-falling process is more stable compared with solution electrospinning. Furthermore, the fluid jet experiences the fastest thinning in the initial region. After that, the stretching process is relatively mild.

5.2.2.4 Effects of the cycle on the jet diameter

The jet diameter that we measured was the last stage of the jet flying, which was just above the collector or the bottom of the simulation system. The average

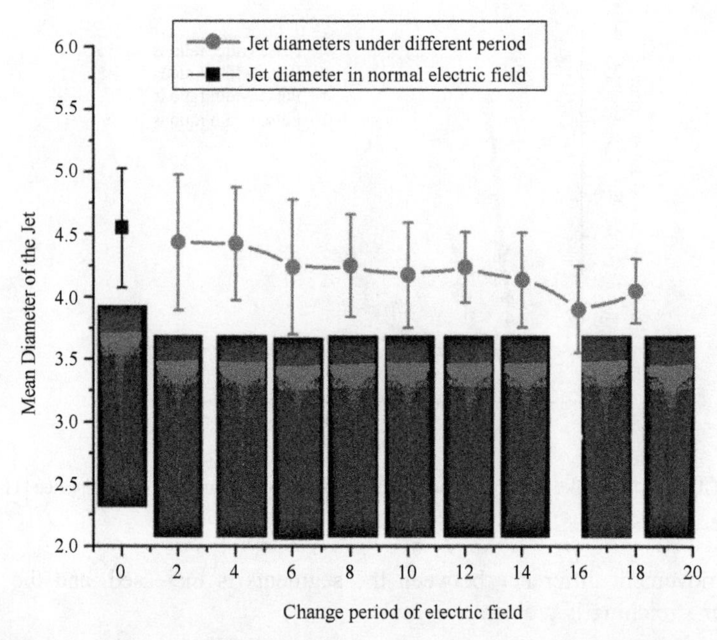

FIGURE 5.13 Changes to the jet diameter in a square wave pulsed electric field with different cycles.

diameter was calculated from the same stage of five repeated simulations of one situation. The repeat was adopted to get a statistical result because of the DPD simulation including a random force [12,32]. As shown in Fig. 5.13, the jet diameter under different cycles in the pulsed electric field is smaller than that in the normal electric field. Each data point is calculated at least 10 times. This proved that the pulsed electric field can reduce the melt electrospinning jet diameter. The reason for this reduction is that the molecular chains in the pulsed electric field have like charge. Hence, adjacent chains and segments repel. The forces among molecular chains change, which makes disentanglement happen easily, so we can get a finer jet. The jet diameter has no obvious change with the increase of the cycle, but cycle 8 obtains the smallest jet diameter. We believe that the reason for this is that our understanding of the mechanisms for charge transport in different cycles of changing electric field is inadequate. However, it may relate to the variation of the electric field. At least, cycle 8 is more fit for the polymer quantity than that used for fiber tensile and the force, changing times as a result.

5.2.2.5 Effect of the pulse width of the square wave on the jet diameter

As mentioned above, the continuous action time of the pulsed electric field was defined as the pulse width. A longer pulse width with the same period

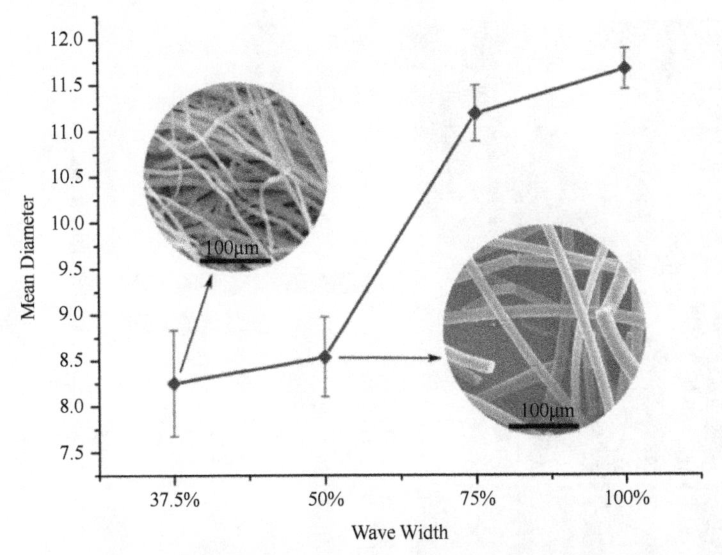

FIGURE 5.14 Diameter of fibers in different duty cycles by DPD (The enlarged images show the SEM at the wave width ratios: 37% and 49% respectively, and the limiting value of the aspect ratio in the experiment is 50%) [11].

(larger duty cycle) results in a larger average electric field force. The effect of the duty cycle on jet diameter was studied.

The jet diameter was calculated using the previous method. The result (Fig. 5.14) revealed that the jet diameter increases as the duty cycle increases. We found that the jet diameter experienced a sharper increase when the duty cycle increased from 50% to 75%. This indicates that the 50% pulse width is an important parameter for determining the jet diameter. This may be due to the larger average electric field force, which led to greater polymer stretch at the same time as increasing the tensile force. This made the single jet diameter larger.

5.2.2.6 Duty cycle effect on melt electrospinning of PLA fibers

In melt spinning, all other factors being the same, the jet diameter largely determines the fiber diameter. Measurement of the diameter of melt fibers collected under different duty cycles is an easy method to understand the diameter trend. A pulsed power supply that is similar to the intermittent electric field in melt electrospinning is used in the duty cycle experiment. The experimental pulse width is equivalent to the pulse width in simulation, with values of 37.1%, 39.4%, and 49.8% (Fig. 5.14).

The PLA is dried for 4 h in the oven before electrospinning. The frequency of the pulse power supply is 10,000 Hz, the voltage is 40 kV, and the distance between the nozzle and collection plate is 8 cm. However, the difficulty with this procedure is that the flow speed is not fixed. Each experiment is carried out at a free flow speed. From Fig. 5.15, we see that the fiber

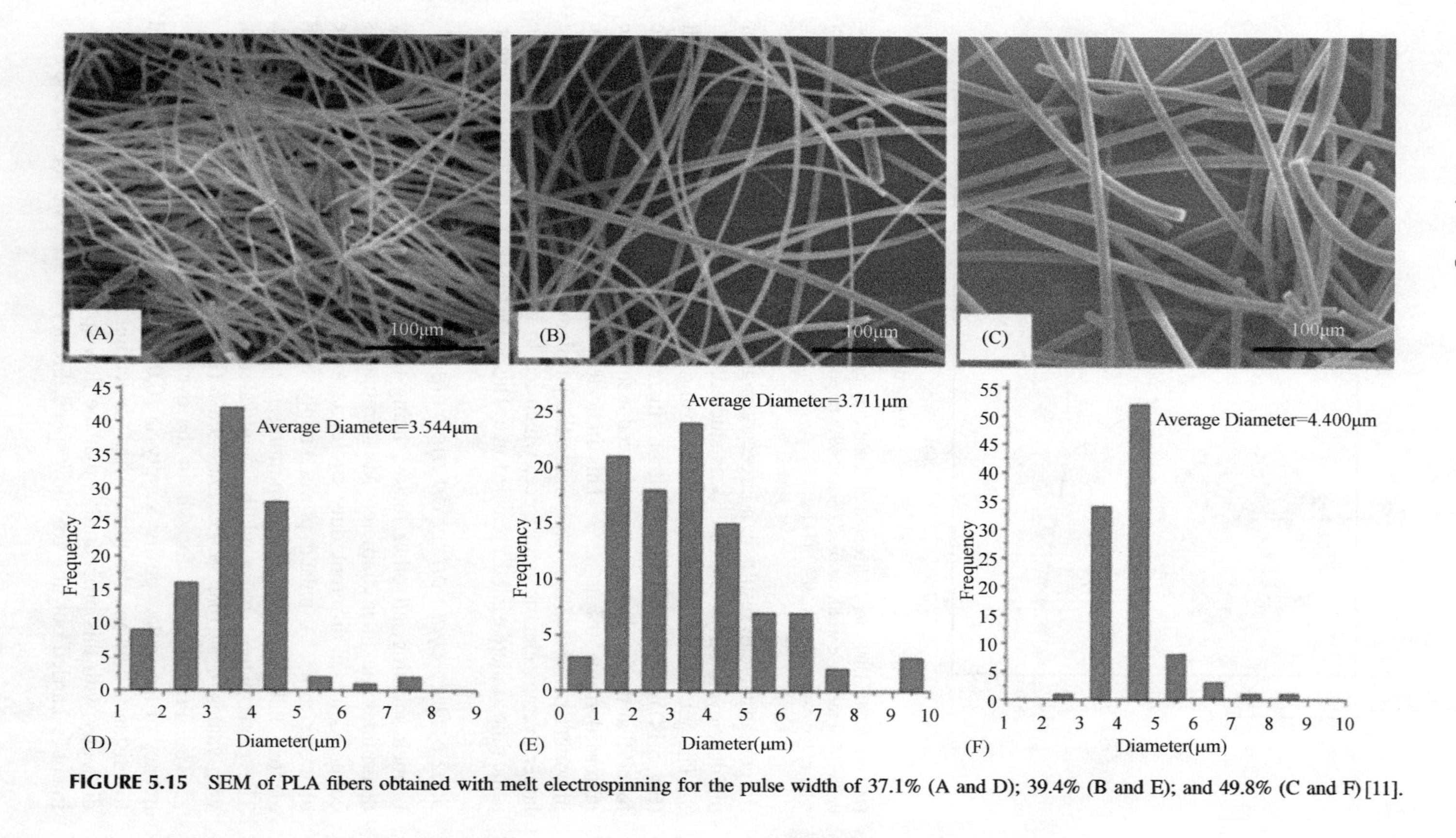

FIGURE 5.15 SEM of PLA fibers obtained with melt electrospinning for the pulse width of 37.1% (A and D); 39.4% (B and E); and 49.8% (C and F) [11].

diameter increases as the duty cycle increases. The diameter distribution becomes narrower when the duty cycle is 49.8%. The fiber surface is smooth at first, but when the duty cycle is larger, fiber fracture occurs. Overall, in the pulsed power supply experiment, the fiber diameter trend with the duty cycle is the same as that in the simulation; the theoretical predictions by DPD simulation are validated with experimental data, indicating that the simulation is a good guide to the melt electrospinning experiments and production.

The pulsed electric field is introduced in the melt electrospinning process. Its effects on molecular stretching and jet diameter of the melt electrospinning process were explored by the DPD simulation method. The molecular stretching is increased in the pulsed electric field, but its cycle does not have an obvious effect on the molecular chain stretching. However, the increase in pulse width can enhance molecular chain stretching. The pulsed electric field decreases the jet diameter effectively and cycle 8 corresponded with the smallest jet diameter. The diameter increased as the pulse width broadened, but they are far less than with the amount of stretching in full pulse width. Finally, an experiment on pulse width is performed, and the fiber diameter and distribution are calculated. We found that, at first, the fiber surface is smooth and its diameter distribution is more concentrated, but when the pulse width was larger, fiber fracture occurred. The fiber diameter change tendency is consistent with the simulation results, which indicates that the simulation can predict and guide the melt electrospinning, which is important for the melt electrospinning process, as well as for the quality of melt electrospinning fibers.

5.3 Conclusion

Dissipative particle dynamics provides new ideas for broadening the research field of electrospinning. The DPD simulation can reveal the molecular stretching process, which cannot be observed using conventional testing methods. The mesoscale simulation method has been used to study the downward traces of melt electrospinning, the drop velocity, the structure of fiber and the chain length, the polymer viscosity, the spring coefficient, etc. It can also verify the simulation results through experimental conclusions, which has irreplaceable advantages.

References

[1] Dzenis Y. Material science: spinning continuous fibers for nanotechnology. Science 2004;304(5679):1917–9.

[2] Hunley MT, Long TE. Electrospinning functional nanoscale fibers: a perspective for the future. Polymer International 2008;57(3):385–9.

[3] Miyauchi M, Miao J, Simmons TJ, et al. Conductive cable fibers with insulating surface prepared by coaxial electrospinning of multiwalled nanotubes and cellulose. Biomacromolecules 2010;11(9):2440–5.

[4] Baumgarten PK. Electrostatic spinning of acrylic microfibers. Journal of Colloid and Interface Science 1971;36(1):71–9.

[5] Hardick O, Stevens B, Bracewell DG. Nanofibre fabrication in a temperature and humidity controlled environment for improved fibre consistency. Journal of Materials Science 2011;46(11):3890–8.

[6] Hutmacher DW, Dalton PD. Melt electrospinning. Chemistry - An Asian Journal 2011;6:44–56.

[7] Huang ZM, Zhang YZ, Kotaki M, et al. A review on polymer nanofibers by electrospinning and their applications in nanocomposites. Composites Science and Technology 2003;63(15):2223–53.

[8] Sill TJ, Recum HAV. Electrospinning: applications in drug delivery and tissue engineering. Biomaterials 2008;29(13):1989–2006.

[9] Zhou FL, Gong RH, Porat I. Three-jet electrospinning using a flat spinneret. Journal of Materials Science 2009;44(20):5501–8.

[10] Groot RD, Warren PB. Dissipative particle dynamics: bridging the gap between atomistic and mesoscopic simulation. The Journal of Chemical Physics 1997;107(11):4423–35.

[11] Song QS, mesoscopic simulation and experimental study on melt electrospinning process in the changing electric fields, Beijing University of Chemical Technology, 2016.

[12] Yong L, Kong B, Yang X. Studies on some factors influencing the interfacial tension measurement of polymers. Polymer 2005;46(8):2811–6.

[13] Yong L, Yang X, Yang M, et al. Mesoscale simulation on the shape evolution of polymer drop and initial geometry influence. Polymer 2004;45(20):6985–91.

[14] Freire JJ, Rubio AM. Conformational properties and Rouse dynamics of different dendrimer molecules. Polymer 2008;49(11):2762–9.

[15] Ibergay C, Malfreyt P, Tildesley DJ. Electrostatic interactions in dissipative particle dynamics: toward a mesoscale modeling of the polyelectrolyte brushes. Journal of Chemical Theory and Computation 2009;5(12):3245–59.

[16] Liu Y, Wang X, Yan H, et al. Dissipative particle dynamics simulation on the fiber dropping process of melt electrospinning. Journal of Materials Science 2011;46(24):7877–82.

[17] Liu Y, An Y, Yan H, et al. Influences of three kinds of springs on the retraction of a polymer ellipsoid in dissipative particle dynamics simulation. Journal of Polymer Science Part B: Polymer Physics 2010;48(23):2484–9.

[18] Groot RD. Electrostatic interactions in dissipative particle dynamics—simulation of polyelectrolytes and anionic surfactants. The Journal of Chemical Physics 2003;118(24): 11265–77.

[19] González-Melchor M, Mayoral E, Velázquez ME, et al. Electrostatic interactions in dissipative particle dynamics using the Ewald sums. The Journal of Chemical Physics 2006;125(22):253.

[20] Duan HW. Design of high-voltage electrospinning machine and optimizing of the electrostatic field (Ph.D. thesis). China: Northeast Forestry University; 2008.

[21] Service RF. Ever-smaller lasers pave the way for data highways made of light. Science 2010;328(5980):810–1.

[22] Liu Y, Deng R, Hao M, et al. Orthogonal design study on factors affecting on fibers diameter of melt electrospinning. Polymer Engineering and Science 2010;50(10):2074–8.

[23] Zhmayev E, Cho D, Yong LJ. Nanofibers from gas-assisted polymer melt electrospinning. Polymer 2010;51(18):4140–4.

[24] Chaudhri A, Lukes JR. Velocity and stress autocorrelation decay in isothermal dissipative

particle dynamics. Physical Review E - Statistical, Nonlinear and Soft Matter Physics 2010; 81(2):026707.

[25] Abunada E. Heat transfer simulation using energy conservative dissipative particle dynamics. Molecular Simulation 2010;36(5):382–90.

[26] Thompson CJ, Chase GG, Yarin AL, et al. Effects of parameters on nanofiber diameter determined from electrospinning model. Polymer 2007;48(23):6913–22.

[27] Yeo LY, Gagnon Z, Chang HC. AC electrospray biomaterials synthesis. Biomaterials 2005;26(31):6122–8.

[28] Pokorny P, Kostakova E, Sanetrnik F, et al. Effective AC needleless and collectorless electrospinning for yarn production. Physical Chemistry Chemical Physics 2014;16(48): 26816–22.

[29] Li C, Su D, Su Z, et al. Fabrication of multiwalled carbon nanotube/polypropylene conductive fibrous membranes by melt electrospinning. Industrial and Engineering Chemistry Research 2014;53(6):2308–17.

[30] Li C, Mu D, Zhang A, et al. Morphologies and crystal structures of styrene–acrylonitrile/isotactic polypropylene ultrafine fibers fabricated by melt electrospinning. Polymer Engineering and Science 2013;53(12):2674–82.

[31] Singer JC, Ringk A, Giesa R, et al. Melt electrospinning of small molecules. Macromolecular Materials and Engineering 2015;300(3):259–76.

[32] Zhang Q, Liu H, Zhu Q, et al. Patterns and functional implications of platelets upon tumor "education". The International Journal of Biochemistry and Cell Biology 2017;90:68–80.

Chapter 6

Experimental study on centrifugal melt electrospinning

Chapter outline

6.1 Overview of centrifugal melt electrospinning

Centrifugal spinning and electrospinning are the traditional methods of making fibers. Electrostatic spinning equipment is simple and safe to operate. It can use solution or melts as the raw material and can prepare ultrafine fibers with good performance. The disadvantage is that it cannot produce ordered fibers, and the spinning efficiency is low, which limits the development of its industrialization [1,2]. In centrifugal spinning, the rotating head consists of polymer melt; the centrifugal force produced by the rotating head breaks the surface tension of the polymer melt and stretches the fiber. Due to the need for high-speed rotation, there is a level of risk with equipment operation.

Melt Electrospinning. https://doi.org/10.1016/B978-0-12-816220-0.00006-3

The advantages are that it is low cost, there is no requirement for a high-voltage electric field, high-spinning efficiency, and ordered fiber membranes can be prepared. Fibers fabricated by centrifugal spinning range in size from micro to nano diameters, but at the same time the mechanical properties of the whole fiber cannot guarantee that it is uniform; the fiber morphology also has significant differences. Therefore, the fibers prepared by centrifugal spinning cannot be applied to high-standard fields [3–5]. Centrifugal electrospinning is a combination of centrifugal spinning and electrospinning, using low voltage and low speed to spin [6]. The simultaneous effect of centrifugal force and electrostatic force is to eliminate the bending instability of the fiber jet only in the electrostatic field, so that the order of the prepared fiber and the mechanical properties are better, as shown in Fig. 6.1.

Depending on the spinning raw materials, centrifugal electrospinning is divided into centrifugal melt electrospinning and centrifugal solution electrospinning. Compared with centrifugal solution electrospinning, centrifugal melt electrospinning can prepare green nanofibers without solvent, which not only saves time in the preparation of the solution but also saves the cost of solvent. At the same time, compared with traditional melt electrospinning, high-efficiency centrifugal melt electrospinning can prepare highly ordered fiber membranes, also ensures the performance of the fiber, which can be adjusted to promote the practical application of nanofibers. However, the centrifugal melt electrospinning machine also combines the three factors of voltage, temperature, and speed. The more influential factors make the spinning process more complicated, and the equipment is more difficult to control. The spinning unit contains more components and is more structural than the normal two ways.

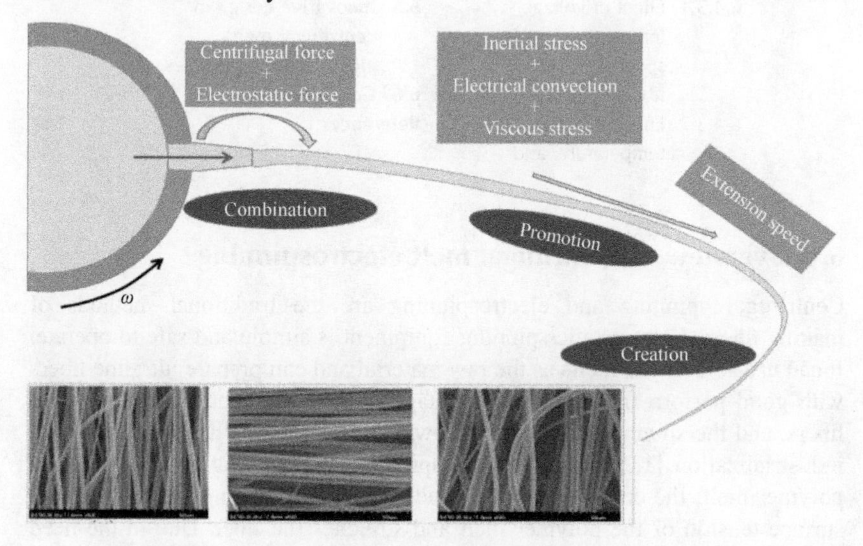

FIGURE 6.1 Schematic diagram of centrifugal electrospinning jet flow [6].

Therefore, the design and layout of the device need to be more rational and precise.

6.2 Research progress of centrifugal melt electrospinning at home and abroad

As shown in Fig. 6.2, the first centrifugal electrospinning device was reported in 2005 by Andrady et al. with US patent 7134857. They invented an electrospinning device with a rotatable spray head [7]. As shown in Fig. 6.2C, in 2012, the centrifugal melt electrospinning method appeared in an article by Li et al. They were spinning in a cavity doped with nickel nanoparticles to successfully produce fibers with diameters of 2–50 μm [9]. The first illustration of the centrifugal solution electrospinning device [10]was shown in Fig. 6.2B. In 2010, Liao et al. were the first to study the degree of crystallinity of bisphenol polycarbonate fibers produced by centrifugal solution electrospinning. They reported fiber prepared by electrospinning had mechanical properties that surpassed the conventional process [8,11]. In recent years, the development of centrifugal electrospinning technology has been very rapid, although the total number of articles on centrifugal electrospinning is low, relevant patent reports are constantly being filed.

In addition, in the field of the development of centrifugal electrospinning, we can summarize centrifugal solution electrospinning and centrifugal melt electrospinning articles in chronological order. Regarding the history of the development of centrifugal solution electrospinning: in 2010, Li et al. prepared a metal mesh in a rotating cylinder, and the holes of the metal mesh replaced the traditional spinning nozzle. This prevented the nozzle from clogging. Therefore, the equipment was very efficient and could be used for large-scale industrial production [12]. One year later, in an article published by Li et al. a 90° angled nozzle was used for electrospinning, and the collecting device was located at the lower end of the spinning nozzle. The fibers were collected in a novel vertical direction. Applying this device, not only did the fiber membrane have an excellent orientation, but also a plurality of intersecting two-layer film could be prepared. The most important thing was that the spinning process of the equipment was simple and safe [13]. In 2012, Valipour designed a gas-sealed centrifugal electrospinning device. The airflow in centrifugal electrospinning has always been one of the important factors that interfere with fiber morphology. In response to this problem, they fixed an insulating airflow plate on the spinning cylinder of the device, so that the rotating device and the collecting device were synchronously rotated in a relatively stationary manner, and the airflow was isolated from the device. Therefore, the ordered degree and the mechanical properties of the overall fiber were better. In addition, the isolated gas flow reduced evaporation of the solvent. This also applied to materials with high solvent evaporation rates, which are important for the development of centrifugal spinning [14]. In the same year, Dabirian et al.

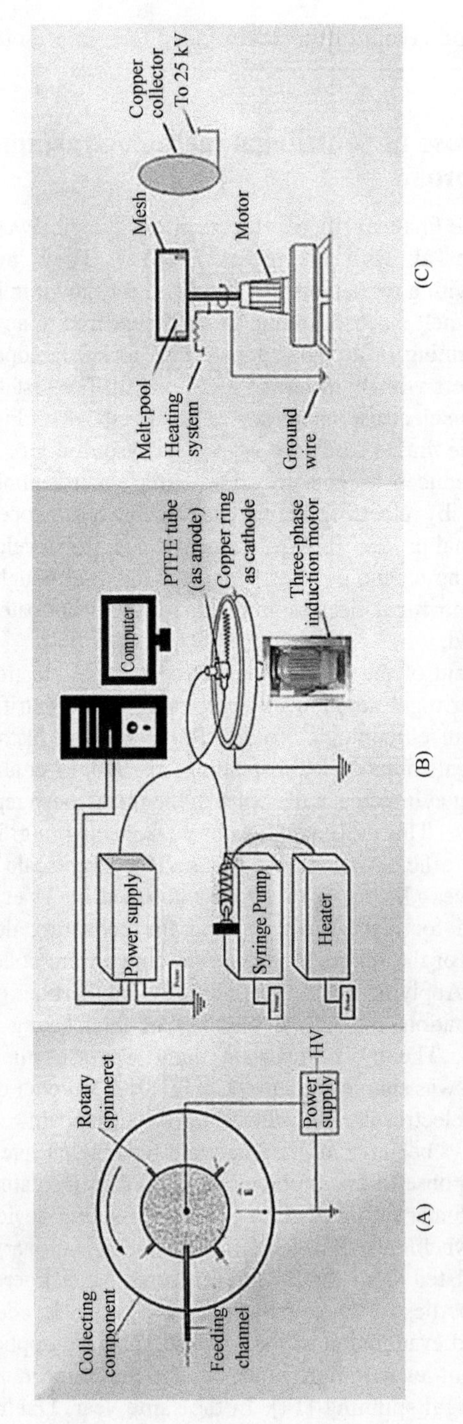

FIGURE 6.2 Schematics of different centrifugal electrospinning systems. (A) The first centrifugal electrospinning device in patent [7], (B) the first centrifugal solution electrospinning device in article [8], (C) the first centrifugal melt electrospinning device [9].

constructed a vertical centrifugal electrospinning device with horizontally arranged rotating axes and a vertical collection [15]. However, the device designed by Edmondson et al. distributed the collecting plates of the parallel electrodes evenly in a circumferential direction. The prepared fiber had a high-order degree, uniform overall performance, and a large specific surface area [16]. In patent 102828260A, Liu et al. designed a coaxial spinning device consisting of an inner cylindrical roller with a stud and an outer cylinder. The spinning holes were one to three rows of small holes uniformly arranged on the upper surface of the outer cylinder. A large number of spinning pinholes and good sealing on the device enabled the scale and continuous production of nanofibers [17]. By 2013, Liu et al. published a report on the spinning equipment for adding T-frame collection parts. The role of the T-frame was to serve as a collection bracket, which could be rotated to obtain a fiber bundle with a stronger mechanically stranded structure [18,19]. In 2014, Xu et al. arranged a plurality of collecting plates in a ring shape around the electrospinning machine. This oriented nanofibers while ensuring spinning efficiency [20].

Regarding the history of centrifugal melt electrospinning development, with the exception of the single article published, only patent-related reports are now available. In patent 103215664A, the device invented by Liu et al. has the following characteristics: the infrared lamp is used for infrared heating, the small extruder is used as the feeding component, and the wired temperature-measuring thermocouple is used as the temperature-measuring component. In addition, the internal structure is a streamlined storage component to ensure no waste of materials [21]. In 2013, Liu et al. designed a new device with a disk with multiple pinhole nozzles as the spinning part. The cross-section of the disk is designed to be upwardly curved, which ensures that the melt material flows sufficiently in the disk chamber without accumulation. The above two devices have the advantages of high spinning efficiency and no blocking [22]. In the same year, in patent 103409861A, Zhang et al. showed a completely new device design. A new feature was the addition of an air compressor and a cyclone vacuum generator to create a swirling airflow within the chamber. The gas stream was prepared by guiding the fibers to achieve a controllable orientation of the fibers to produce a twisted fiber [23,24]. By 2014, in patent 104178826A, the device designed by Zou et al. used a composite structure of the storage member. A channel inside the spinning part branched into a plurality of channels to play the role of cutting the molten material in advance, which helped to achieve efficient spinning, but at the same time, the complicated internal flow passages were difficult to clean [25]. The device constructed by Liu et al. in patent 104088024A used a wireless infrared thermometer, an electromagnetic heating device, and an umbrella-shaped nozzle. The device had the characteristics of real-time temperature measurement and temperature control, and the other two coaxial conical nozzles ensured the efficiency of spinning [26].

In Section 2.8 of Chapter 2, the new generation of centrifugal melt electrospinning equipment designed by the author's research group has been described in detail and will not be repeated here.

6.3 The significance of centrifugal melt electrospinning devices

At present, the traditional spinning method has some disadvantages that cannot be overcome, and the conventional electrospinning method is inefficient. Although needleless, multineedle, and umbrella nozzles have been significantly improved, the overall performance and orderliness of the fibers prepared by these devices have to be unified and improved. These shortcomings hinder the industrialization of electrospinning. The centrifugal spinning can efficiently prepare the ordered fiber membranes, but the overall morphology and mechanical properties of the prepared fibers are not uniform and there is a big difference between them. This also limits the application of the fiber membranes in many demanding areas. Both traditional spinning methods have the above-mentioned shortcomings, and currently, as the nanofiber industry market demand continues to increase, by combining the advantages of the two methods, the centrifugal electrostatic spinning method has been developed.

Although centrifugal solution electrostatic spinning has been in rapid development for nearly a decade, this method can only use solution for spinning. The presence of residual solvents in the prepared fibers is a serious problem in the area of high demand for medical organisms. To overcome this problem, the centrifugal melt electrospinning method has been developed. The centrifugal melt electrospinning device is more complex than other methods of spinning and risks will be relatively increased. However, the centrifugal melt electrospinning method has some significant advantages. Centrifugal melt electrospinning can be directly used with solid melt material, and has high spinning efficiency. The preparation of the fiber membrane is completely solvent-free, is guaranteed to have no side effects, and the performance of the fiber membrane can be controlled by adjusting the process parameters. Therefore, the centrifugal melt electrospinning technology and fiber preparation industry are very consistent with the development trend of industrialization. The centrifugal melt electrospinning device has been self-designed by various researchers, and therefore, undergoes several experiments and simulation studies to prepare greener nanofibers, which are applicable in various fields, especially in biomedical applications.

There are specific problems with the centrifugal melt electrospinning approach. First, there is no complete set of defective centrifugal melt electrospinning devices. There are some problems with the temperature parts in the existing centrifugal melt electrospinning equipment. Wireless infrared temperature measurements and wired thermocouple contact temperature measurements contain certain temperature errors. And in the spinning parts, the cable thermocouple temperature will appear to break during rotation. Second, centrifugal electrostatic spinning parameters and external factors are complex. Third, the centrifugal melt electrospinning device is designed by a combination of high temperature, high pressure, and high speed. This therefore leads to some risk factors. Fourth, in the centrifugal melt electrospinning process, the molten material releases odorous gas, which is also one of the important risk factors during the experiment. Fifth, centrifugal electrospinning for the preparation of nanofibers depends upon the application, so we need to continue the design process.

6.4 Experimental study on centrifugal melt electrospinning

In order to explore smooth spinning and good fiber properties, we independently designed a centrifugal melt electrospinning device in our laboratory to prepare nanofibers with solvent-free solid poly L-lactic acid (PLLA) materials. Based on the characterization and analysis of PLLA fiber membranes, the regularity of PLLA fiber membranes was summarized. In addition, the orderly micro-/nanofiber membrane was prepared by centrifugal melt electrospinning technology, and the order degree of fiber membrane was analyzed and characterized. Then those fibers were applied in biomedical experiments. It was verified that the solvent-free PLLA fiber membrane prepared by the centrifugal melt electrospinning method was nontoxic and effective as a biological scaffold. Finally, in order to improve the mechanical properties of PLLA fiber membranes, we tried to explore centrifugal melt electrospinning technology to prepare composite fiber membranes containing PLLA and polyethylene oxide (PEO) and then carried out the characterization and mechanical properties. Compared with the fiber properties of pure PLLA, it was proved that this device could prepare high-performance composite fiber membrane.

6.4.1 Experimental section

Material: L-polylactic acid (PLLA), purchased from Haizheng Biomaterials Co. Ltd., $T_g = 61°C$, $T_m = 171° C$, $M_n = 100{,}000\ g \cdot mol^{-1}$, melt flow rate of $3–5\ g \cdot (10\ min)^{-1}$ (190°C, 2.16 kg).

The centrifugal melt electrospinning device consists of the following five parts (according to each functional part of the mutual cooperation between the group): (1) high-pressure generator, collection ring; (2) inverter, drive motor;

(3) temperature controller, electromagnetic heating device, temperature stent, cable thermocouple; (4) spinning cylinder, umbrella nozzle; and (5) fixed platform, lifting platform. The main functions of each part are as follows:

(1) The first part: the positive pole of the high-voltage generator is directly connected to the collecting ring; the negative pole of the high-voltage generator is connected to the spinning cylinder, which is grounded. On application of a high-voltage supply, an electric field is generated between the spinning cylinder and collecting ring.
(2) The second part: the driving motor is connected to the rotating cylinder through the coupling, and the frequency converter controls the rotation speed of the driving motor, thereby directly controlling the rotation speed of the spinning cylinder.
(3) The third part: the temperature control instrument, the electromagnetic heating device, and the wired thermocouple coordinate to adjust the spinning temperature. The electromagnetic heating device is composed of an electromagnetic heating ring and an exhaust fan, which is located at the lower end of the spinning cylinder, to effectively and uniformly heat the spinning cylinder. The exhaust fan removes the heated gas in time and is used to quickly lower the temperature of the rotating cylinder. The wired thermocouple fixed by the temperature-measuring bracket feeds back the internal temperature of the spinning cylinder. The temperature controller controls the electromagnetic heating coil in the form of Proportion Integration Differentiation (PID) control system to adjust the spinning temperature in real time.
(4) The fourth part: two umbrella nozzles are attached to the side walls of the spinning cylinder. The umbrella nozzles ensure the high-efficiency spinning of the device.
(5) The fifth part: the function of the fixed table is to fix the whole set of devices and the lifting table is used to regulate the height of the thermocouple bracket.

First, as shown in Fig. 6.3, set up a centrifugal melt electrospinning experimental device. At the beginning of the experiment, the temperature is set in the temperature control box (200–300°C), according to the melting temperature of the material. The electromagnetic heating device starts to heat the rotating cylinder and the thermocouple feeds back the internal temperature of the rotating cylinder and reaches the set temperature. After reaching the set temperature, the materials are added into the rotating cylinder and proceed until the spinning material is completely melted. The equipment, such as the collecting plate, is designed to be kept at a distance of 6–12 cm. Then set the high-voltage power supply value (0–50 kV) and speed above the inverter (0–1800 $r \cdot min^{-1}$), to start spinning. Under the influence of centrifugal force, the spinning process occurs.

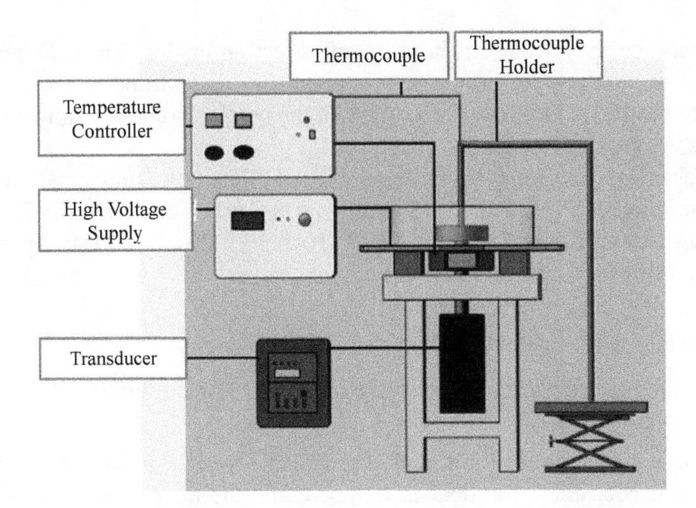

FIGURE 6.3 Schematic diagram of the centrifugal melt electrospinning device [27].

In order to eliminate experimental error and the need for repeating the number of experiments, the double variable method was used for the comparative analysis. The time for collecting the spinning was 30 s, and the weight of the collected fibers was measured with high-precision electrons.

6.4.2 Characterization method

Observation of fiber diameter and morphology is carried out using scanning electron microscopy (SEM) equipment (Hitachi S4700, Japan). All fibrous membranes were subjected to spray treatment prior to observation, and the observed scanning electron microscopy images were analyzed, using software Image J and software Origin 8.0, for fiber diameter.

Observed fiber crystallinity: X-ray diffraction (XRD), apparatus: D/Max 2500 VB2 + /PC (Rigaku, Japan). The XRD data collection range is 2θ at 5–50° and the sweep rate is $3° \cdot min^{-1}$, 0.02. Data analysis for results used Software MDI jade 5.0 and software Origin 8.0.

Yield test: the use of high-precision electronic data, data analysis, and statistical software Origin 8.0.

Differential scanning calorimetry (DSC): Equipment: US Perkin Elmer Pyris 1, heating range 25–130°C, heating rate of $10°C \cdot min^{-1}$, the environment is nitrogen, a flow rate of $20\ mL \cdot min^{-1}$.

Fiber membrane characterization: Using SEM, and then using fast Fourier transform (FFT), with Image J software to install Oval profile plug-in. The SEM image is first transformed into a spectrum, and then the Oval profile is used to calculate the gray value of the spectrum in the range of 0–360° in the radial direction. The resulting quantified spectral intensity distribution map represents the gray value of the different direction angles. Then, Origin 8.0 software is used to perform spectral processing and fitting of the spectral intensity profile, to facilitate comparison of the height and half-width values of the peak feature [28–34].

Mechanical tensile testing equipment: Electronic universal testing machine, model UTM2502, purchased from Shenzhen Synopsis Technology Co. Ltd. Calculation criteria: GB/T 1040.3–2006. The test speed is 5 mm·min^{-1}, film size to take 2 cm × 5 cm.

6.4.3 Results and discussion

6.4.3.1 Effect of voltage, temperature, and speed on PLLA fiber production

First, the weight, temperature, and speed of the device during the spinning process are controlled, the fiber is weighed, and the fiber yield is calculated simultaneously. By using Origin 8.0 software, the relationship between voltage, temperature and speed, and fiber yield, is shown in Fig. 6.4.

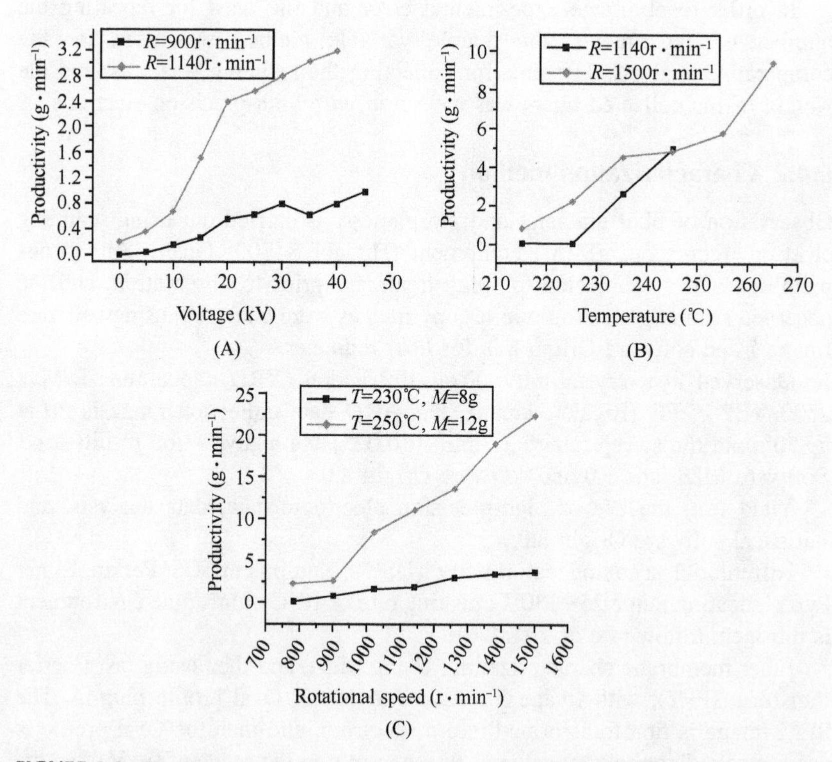

FIGURE 6.4 The summary of kinds of spinning process conditions affecting fiber production. (A) Different voltage and rotating speed: $T = 230°C$, $C = 10$ cm, $M = 8$ g, $V = 0–50$ kV, (1) $R = 900$ r·min^{-1}, (2) $R = 1140$ r·min^{-1}; (B) different temperature and rotating speed: $V = 35$ kV, $C = 10$ cm, $M = 8$ g, $T = 215–265°C$, (1) $R = 1140$ r·min^{-1}, (2) $R = 1500$ r·min^{-1}; (C) different rotating speed and initial mass: $C = 10$ cm, $V = 30$ kV, $R = 780–1500$ r·min^{-1}, (1) $T = 230°C$, $M = 8$ g, (2) $T = 250°C$, $M = 12$ g [27].

Analysis of the above experimental rule: Fig. 6.4A shows that an increase in voltage allows an increase in fiber production. By comparing the curves and the slopes of the two curves, it was concluded that the fiber yield increases obviously with an increase in the rotational speed. The role of speed is stronger than the role of voltage. Fig. 6.4B shows that when the temperature increases, the fiber yield increases significantly. Comparing two curves, when speed is increased, fiber production is also increased. Therefore, fiber production, temperature, and speed have played a significant role. However, after 245°C, with an increase in the rotational speed, the fiber yield does not vary because the initial melting material content is low. It can be concluded from Fig. 6.4C that the initial material content increase there is favorable for the increase in spinning yield. But if the initial content is too great, the centrifugal force becomes greater, the spinning fiber becomes significantly thicker, and this results in the preparation of ultrafine nanofibers not being possible. Analysis obtained from the graph indicates that with the increase in speed, fiber production increased. By contrast, analysis of the curve obtained indicates that at lower temperatures, fiber production increases with slower speed. At higher temperatures, the fiber yield increases faster with speed, and the initial melt material content significantly affects the yield. Under the same conditions, the initial content increased, and fiber production increased more significantly.

In summary, we can state the effect of the process parameters on the fiber yield in the centrifugal melt spinning process. The effect of speed on the fiber yield is that the temperature of the spinning melt first reaches the melting point and melts into a liquid with a certain viscosity. Then, with the increase in rotational speed, the centrifugal force of the spinning raw material is obviously enhanced, and the yield of the fiber increases with the increase in rotational speed.

The effect of voltage on the yield of fiber: the temperature of the spinning melt must first reach the melting point and melt into a liquid with a certain viscosity. Then, with the enhancement of the voltage effect, the electric field force of the spinning raw material is gradually increased, and the yield of the fiber also gradually increases with the increasing voltage.

The effect of temperature on the yield of fiber: when the viscosity of the melt is large, the material will melt as the temperature increases, and the viscosity of the melt gradually decreases, which facilitates better spinning of the material into fibers. With the increase in temperature, the fiber yield increases significantly. But when the temperature is too high, the fiber undergoes a degradation phenomenon, so too high a temperature will lead to decomposition of the fiber. The order of the effect of the process parameters on the fiber yield is: temperature, speed > voltage. The effect of these three process parameters on the yield of fiber is basically proportional to this law.

6.4.3.2 Effect of voltage, temperature, and speed on the fiber diameter of PLLA

The fiber diameter in all SEM images was measured using software Image J, and each fiber was selected with five data points for measurement, and then the data and statistics were analyzed using Origin 8.0 software.

Experimental analysis: in the experimental phenomenon, the higher the voltage, the more significant the increase in the number of fibers reaching the collection circle. From the experimental data analysis, Fig. 6.5 shows that, as the voltage increases, the electric field force in the spinning jet gradually increases. The fibers are sufficiently stretched during the collection process, and the average diameter of the collected ultrafine nanofibers is gradually reduced. Basically, when the voltage increases, the fiber is fully stretched and refined, so

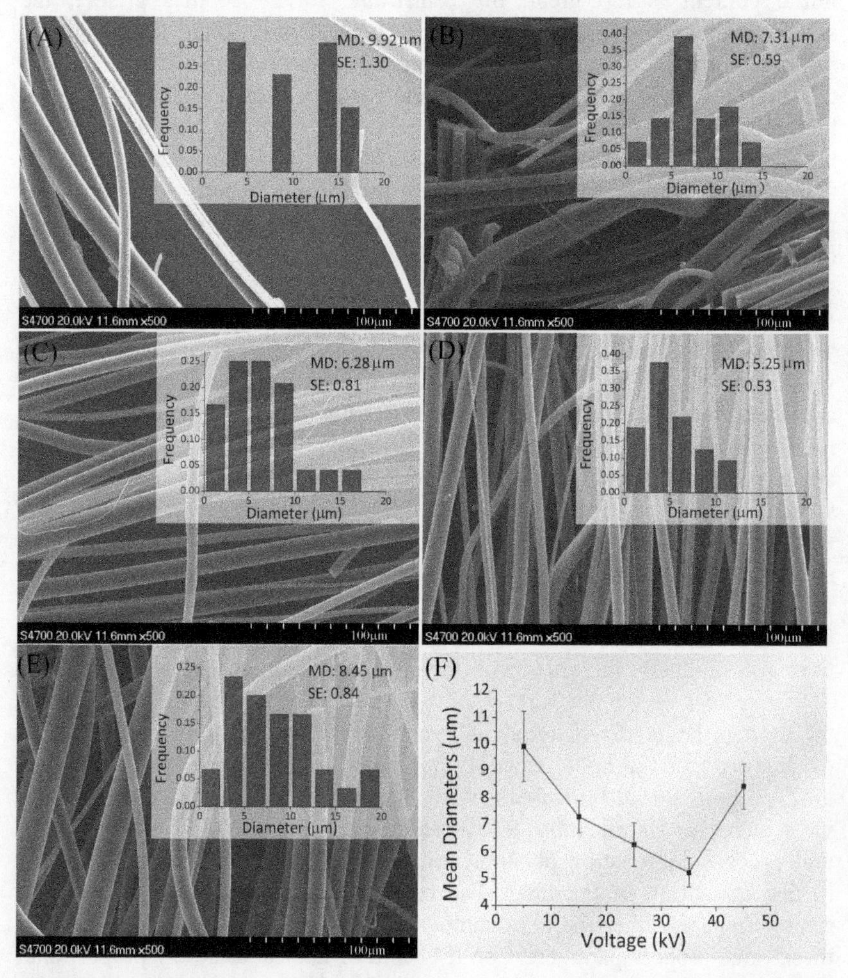

FIGURE 6.5 SEM figures of the diameters of fibers and the tendency chart of the mean diameter of fibers with voltage [27].

that there will be little variance in the fiber diameter. With a further increase in voltage, the jet cannot be fully stretched out and the fiber emerges with thicker diameter and greater breakdown.

Experimental analysis: when the temperature is too low, the spinning jet cannot be formed. As shown in Figs. 6.6 and 6.7, after the temperature of the spinning melt reaches the melting point, the viscosity of the spinning melt gradually decreases as the set temperature gradually increases. The fibers are more easily stretched into ultrafine fibers and are uniformly distributed. But at the same time, the impact of air during rotation causes fluctuation in fiber stretching. Moreover, the higher the spinning temperature, the greater is the temperature difference between the inside and outside of the rotating cylinder, which will weaken the initial jet drawing.

Experimental analysis: As shown in Figs. 6.8 and 6.9, the average diameter of the fiber decreases first and then increases as the speed increases. This is

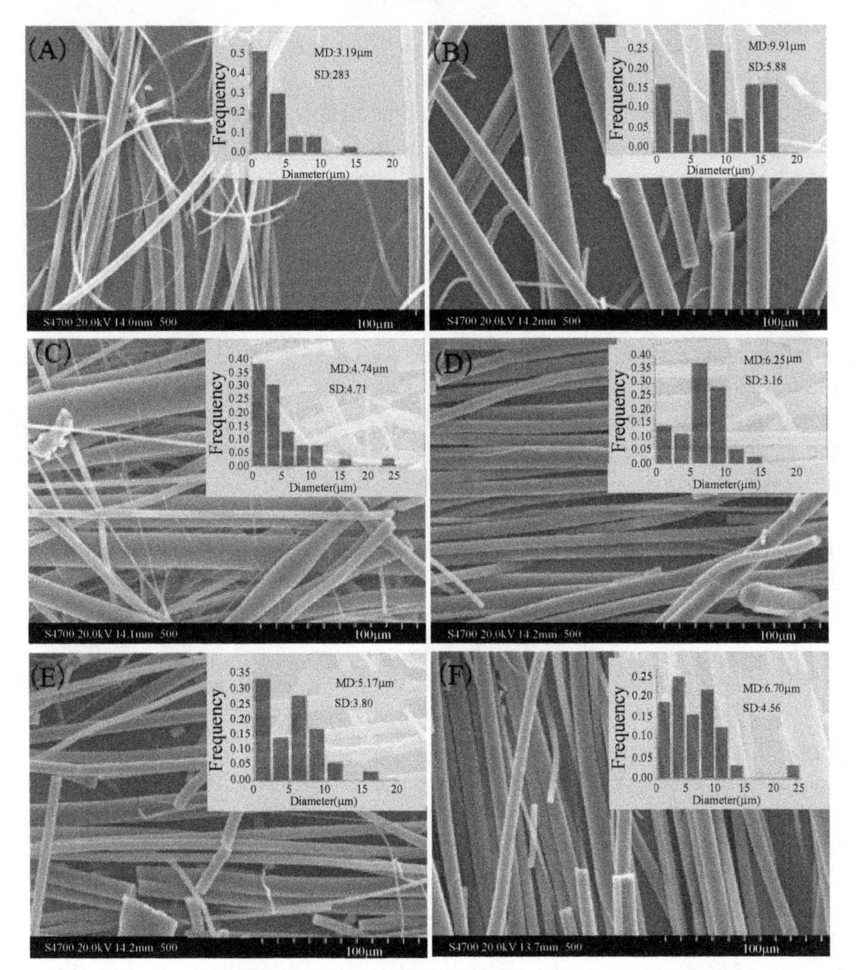

FIGURE 6.6 SEM figures of the diameters of fibers [27].

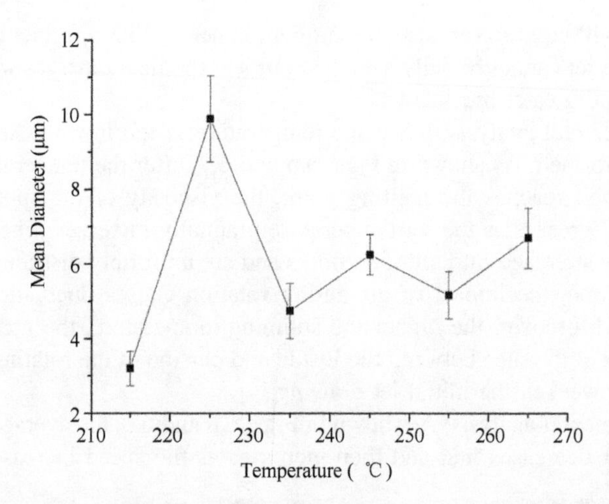

FIGURE 6.7 Tendency chart of the mean diameter of fibers with temperature [27].

because, as the speed increases, the rotating airflow is significantly enhanced, which causes the jet to dry and the jet path to become unstable. Centrifugal force plays a major part in the spinning jet. At low speed, the centrifugal force acts more strongly on the jet, so the fiber diameter is finer. At high speed, the influencing factors of the rotary airflow are more obvious, so the jet is more disturbed by the interference factor, and the fiber diameter becomes thicker.

Experimental rules summary: for fiber diameter effects, the process parameters affect the strength of the order: voltage > speed > temperature. In the centrifugal melt electrospinning process, the collected fiber diameter distribution is wide, from several tens of nanometers to a few microns. To achieve the nano-level entire fiber membrane we also need to continue to explore the process conditions.

Fig. 6.10 shows an SEM image of the mean diameter of ultrafine fibers. As the voltage increases, the production of ultrafine fibers increases. In addition, as shown in Fig. 6.11, it indicates that ultrafine nanofibers are electrospun by centrifugal melt electrospinning with a diameter of 273 nm.

6.4.3.3 *Effect of voltage, temperature, and speed on the crystallinity of PLLA fibers*

Since the PLLA material itself is a slow crystalline polymer, complete crystallization is not possible and fiber crystallinity is low. However, different spinning process conditions have a significant effect on the ordering of molecular chains in the fiber.

The experimental data are analyzed using MDI jade 5.0 software and Origin 8.0 software. Analysis of the data and phenomena in Fig. 6.12A shows that when a voltage is enhanced, the fiber is obviously fully stretched and the number of fibers reaching the collection plate is significantly increased. When the voltage increases, the electric field tensile effect increases, and there is an

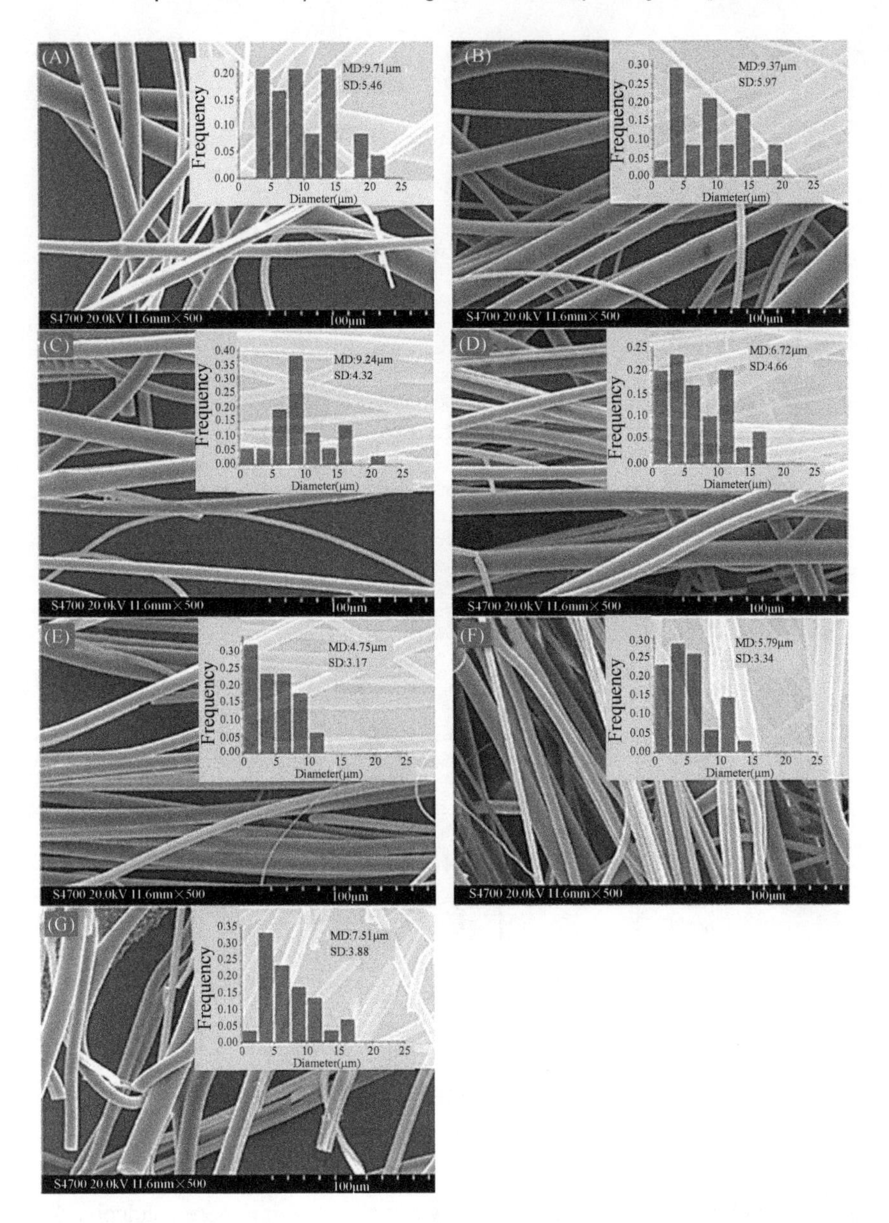

FIGURE 6.8 SEM figures of the diameter of fibers[27].

increase in crystallinity. It is therefore important to note that when the voltage increases beyond 45 kV, the electric field force is unstable and fiber crystallinity decreases.

Analysis of the data in Fig. 6.12B in the centrifugal melt electrospinning process indicated that when the melting temperature increases, pinning

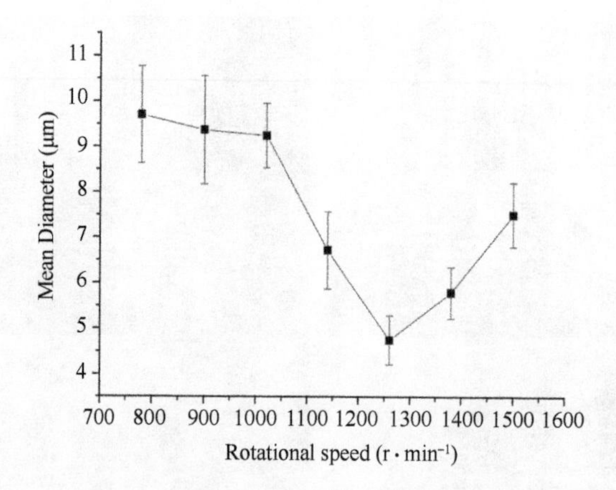

FIGURE 6.9 Tendency chart of the mean diameter of fibers with rotational speed [27].

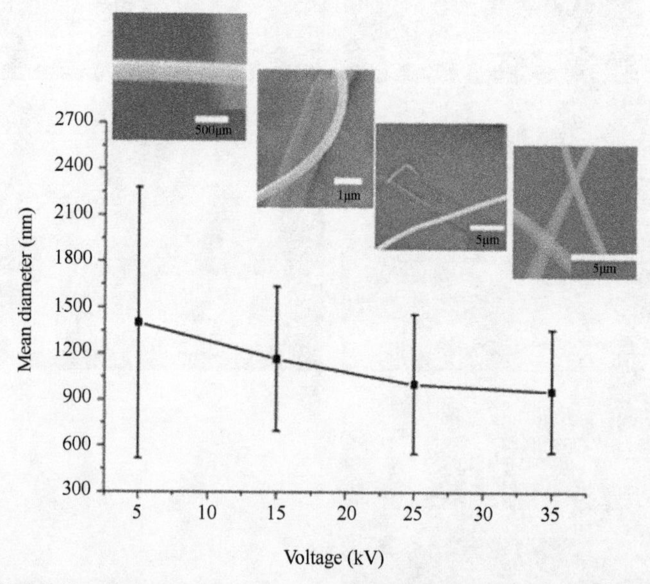

FIGURE 6.10 Tendency chart of the mean diameter of ultrafine fibers with voltage [27].

material melts quickly. Material viscosity decreased, being more conducive to fiber stretching and refinement, which is more conducive to the crystallization process. However, since the crystallization rate of the PLLA material is too slow, the set temperature is high and the ambient temperature is constant. As a result, the curing temperature difference of the jet fiber is large, the fiber is sufficiently stretched, and the fiber crystallization process cannot be completed. Therefore, the law of the influence of temperature on the crystallinity of the fiber cannot obtain regular data.

FIGURE 6.11 SEM figures of ultrafine fibers spun by centrifugal melt electrospinning [27].

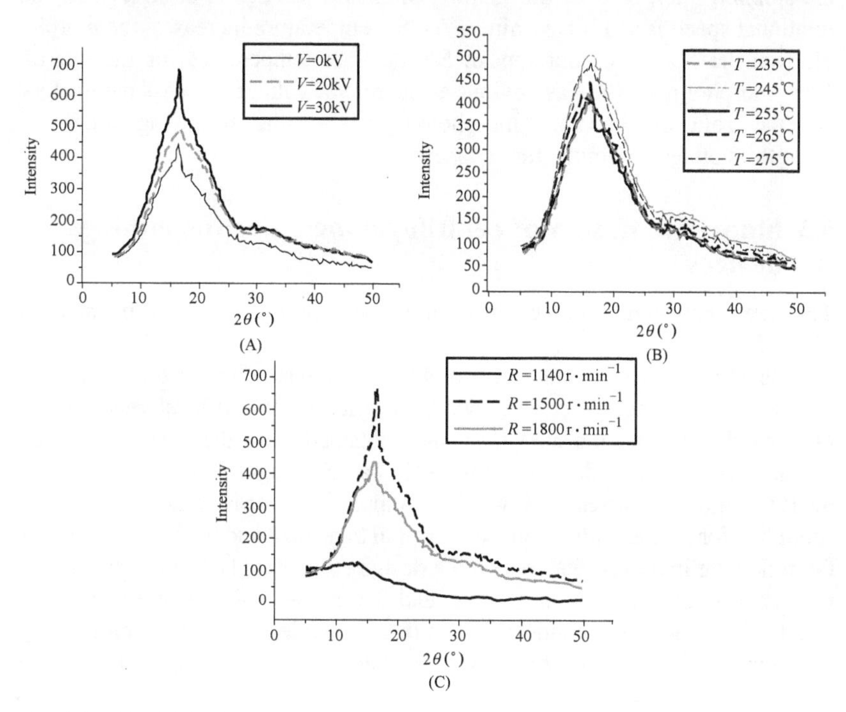

FIGURE 6.12 (A) XRD figure of PLLA fibers in conditions of variable voltage, (B) XRD figure of PLLA fibers in conditions of variable temperature, (C) XRD figure of PLLA fibers in conditions of variable rotational speed [27].

Analysis of the data in Fig. 6.12C shows that as the speed increases, the fiber tensile force increases. In addition, the initial velocity of the fiber jet increases and the time to reach the collecting plate is shortened. At the same time, the surrounding swirling airflow is enhanced and the fiber is accelerated by the drying process. Proper lifting speed is beneficial to the fiber being fully stretched, and the speed is too great to facilitate stretching. Therefore, as the rotation speed increases, the order of the molecular chains in the fibers first increases and then decreases. Eventually, the crystallinity is also first increased and then decreased.

6.4.3.4 *Effect of voltage, temperature, and speed on the morphology of PLLA fibers*

The effect of different process conditions on the morphology of the fiber in centrifugal melt electrospinning has been discussed. With an increase in voltage, there is an increase in the tensile force of the electric field and the fiber shape becomes better. As shown in the figure, the bead structure of the fiber decreases obviously with an increase in voltage.

The effect of temperature on the morphology of the fiber shows that after the spinning melt reaches the melting point, the voltage is at 35 kV and the rotational speed is at 1140 $r \cdot min^{-1}$. As the temperature increases, the morphology of the fiber does not appear. So the fiber temperature for the role of fiber morphology is not as obvious as the role of voltage. Process parameters for the role of the fiber morphology are in the following order of strength: voltage > temperature > speed.

6.5 Innovative design of centrifugal melt electrospinning devices

The new centrifugal melt electrospinning device diagram is illustrated in Fig. 6.13.

The above devices illustrate the problems in the structure of the centrifugal melt electrospinning device designed by our laboratory with reference to the previous design, and the analysis results obtained from the experiments and simulations. The final design of Fig. 6.13, "a centrifugal melt electrospinning device combining wired and wireless temperature measurement" has been submitted for patent application. As shown in Figs. 6.14 and 6.15, the device has the following improved features on the design structure. (1) The temperature is measured by a contact thermocouple, and the thermocouple wireless temperature feedback device is connected. The thermocouple is rotated synchronously with the rotating cylinder by fixing the thermocouple to a special spinning

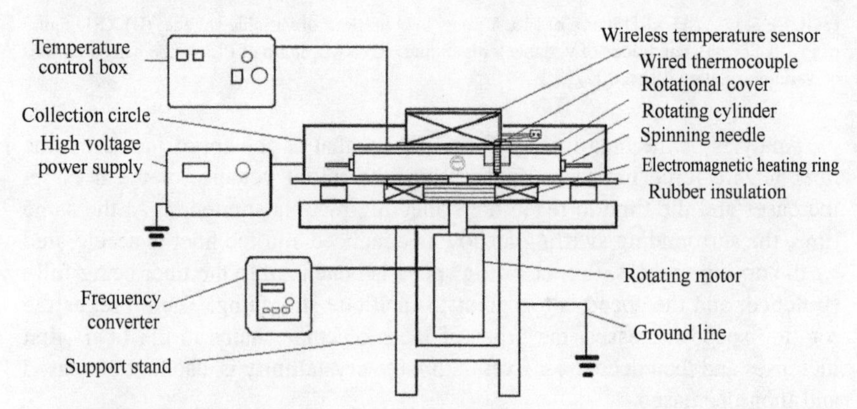

FIGURE 6.13 Schematic diagram of the new centrifugal melt electrospinning device [35].

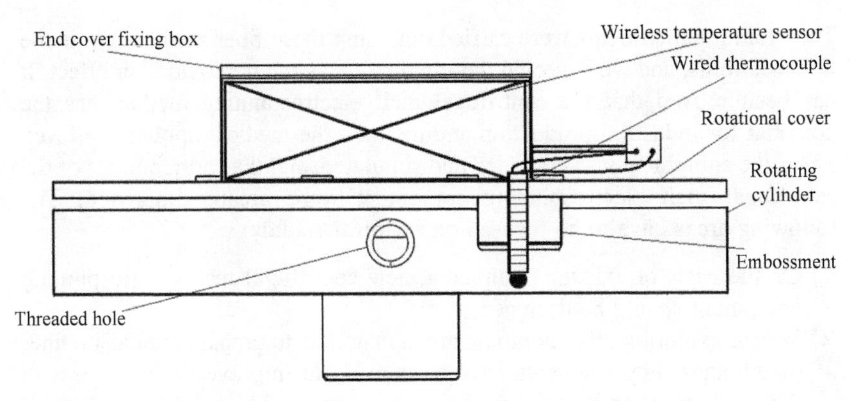

FIGURE 6.14 The rotating cylinder parts of the combination of wireless temperature sensor and thermocouple cable [35].

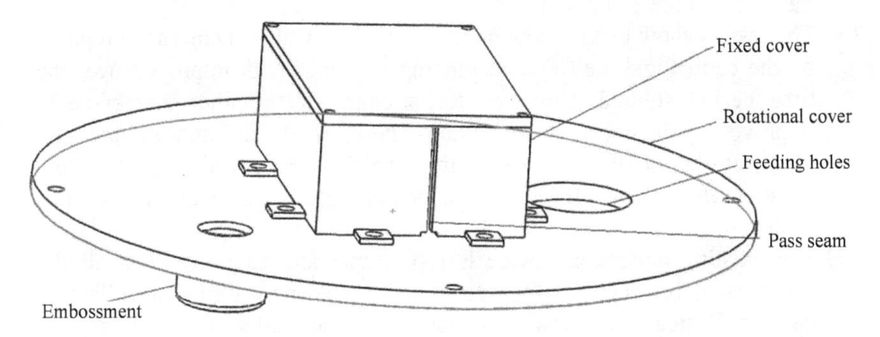

FIGURE 6.15 New rotating cylinder cover [35].

end cap. Therefore, the thermocouple changes from an original dynamic temperature measurement to a static temperature measurement, and the original thermocouple bracket is removed. In this way, there is no safety problem with the thermocouple bracket and the problem of the thermocouple being rubbed and hinged, and the temperature measurement is more accurate. (2) The rotating cylinder is designed in the form of a cross-flow passage with a raised projection. Therefore, the heated area of the material is limited to the circumference of the rotation, the heat is uniform, and there is no temperature difference in the material. (3) Four threaded holes are arranged around the rotating cylinder, and four spinning needles with adjustable diameters are installed to ensure high-efficiency spinning.

6.6 Conclusion

We investigated the effect of detailed process parameters on the performance of PLLA fibers in centrifugal melt electrospinning, and prepared PLLA fiber membranes with different thicknesses and the order degrees were studied.

The biofilm experiments were carried out using these fiber membranes as the fiber scaffolds, and we expected the same effect as the nontoxic fiber effect. It has been proved that the centrifugal melt electrospinning method has the potential for industrial production and to reach the medical application level. Later, the equipment was upgraded and simulated, and the shortcomings of the centrifugal melt electrospinning equipment were greatly improved. The following areas have to be focused on for further study:

(1) On the basis of existing equipment, new centrifugal melt electrospinning equipment should be designed.
(2) We are exploring other nontoxic green materials to prepare composite fiber membranes. Improvements have been made to improve the toughness in the mechanical properties of the fiber membrane. Thus, the application of the fiber membrane prepared by the centrifugal melt electrospinning method is more promising.
(3) The results show that the toughness of the PLLA fiber membrane prepared by the centrifugal melt electrospinning method is not improved over the time period studied. However, the strength of the fiber membrane is improved by the hot pressing method. Therefore, the different hot pressing temperatures for fiber membrane improve the mechanical properties, and there is also a need to conduct a detailed study and summary of this subject.
(4) The existing collection device is too simple and cannot collect all the fibers, so it is necessary to explore and design new forms of collection devices. Basically, the new collection device can collect all the fibers, and can collect different widths of the fiber membrane, laying the foundation for future industrial equipment.

After all equipment upgrades and manufacturing are completed, we must try to prepare fiber membranes with different angles of the cross-structure. The mechanical properties of the cross-structured fiber membranes were explored, and the fiber membranes compared with the ordered structures will be greatly improved.

References

[1] Brown TD, Dalton PD, Hutmacher DW. Melt electrospinning today: an opportune time for an emerging polymer process. Progress in Polymer Science 2016:116–66.

[2] Zhao GX, Zhang XH, Lu TJ, et al. Recent advances in electrospun nanofibrous scaffolds for cardiac tissue engineering. Advanced Functional Materials 2015;25(36):5726–38.

[3] Kamal S, Carlos G, Steve Z, et al. Electrospinning to Forcespinning™. Materials Today 2010;13(11):12–4.

[4] Lu Y, Li Y, Zhang S, et al. Parameter study and characterization for polyacrylonitrile nanofibers fabricated via centrifugal spinning process. European Polymer Journal 2013;49(12):3834–45.

[5] Mceachin Z, Lozano K. Production and characterization of polycaprolactone nanofibers via forcespinningTM technology. Journal of Applied Polymer Science 2012;126(2):473–9.
[6] Ren L, Ozisik R, Kotha SP, et al. Highly efficient fabrication of polymer nanofiber assembly by centrifugal jet spinning: process and characterization. Macromolecules 2015;48(8):2593–602.
[7] Andrady AL, Ensor DS, Newsome JR. Electrospinning of fibers using a rotatable spray head. 2005.
[8] Liao CC, Wang CC, Shih KC, et al. Electrospinning fabrication of partially crystalline bisphenol A polycarbonate nanofibers: effects on conformation, crystallinity, and mechanical properties. European Polymer Journal 2011;47(5):911–24.
[9] Li J, Guo Q, Shi J, et al. Preparation of Ni nanoparticle-doped carbon fibers. Carbon 2012;50(5):2045–7.
[10] Peng H, Zhang JN, Li XH, et al. Modes of centrifugal electrospinning. Engineering Plastics Application 2015;(09):138–42.
[11] Liao CC, Hou SS, Wang CC, et al. Electrospinning fabrication of partially crystalline bisphenol A polycarbonate nanofibers: the effects of molecular motion and conformation in solutions. Polymer 2010;51(13):2887–96.
[12] Li SJ, Li J, Zhang YC, et al. Centrifugal electrospinning device. 2011. CN 102061530 A.
[13] Li M, Long YZ, Yang D, et al. Fabrication of one-dimensional superfine polymer fibers by double-spinning. Journal of Materials Chemistry 2011;21(35):13159–62.
[14] Valipouri A, Hosseini Ravandi SA, Pishevar AR. A novel method for manufacturing nanofibers. Fiber Polymer 2013;14(6):941–9.
[15] Dabirian F, Hosseini Ravandi SA, Pishevar AR. The effects of operating parameters on the fabrication of polyacrylonitrile nanofibers in electro-centrifuge spinning fibers and polymers. Fiber Polymer 2013;14(9):1497–504.
[16] Edmondson D, Ashleigh C, Soumen J, et al. Centrifugal electrospinning of highly aligned polymer nanofibers over a large area. Journal of Materials Chemistry 2012;22:18646–52.
[17] Liu YB, Zhang ZR, Qi DY, et al. Centrifugal needleless electrospinning device. 2012. CN 102828269 A.
[18] Long YZ, Yin HX, Liu SL, et al. A device for preparing nanofiber stranded structure. 2011.10.12. CN 102212893A.
[19] Liu SL, Huang YY, Han YM, et al. Preparation of fluorescent nanofibers with ordered, crossover and stranded structure by centrifugal electrospinning. Journal of Qingdao University (Natural Science Edition) 2013;26(1):45–9.
[20] Kancheva M, et al. Advanced centrifugal electrospinning setup. Materials Letters 2014;136:150–2.
[21] Liu Y, Li XH, Chen ZY, et al. A centrifugal electrostatic spinning device. 2013. CN103215664A.
[22] Liu Y, Song TD, Chen ZY, et al. An intermittent centrifugal melt electrospinning device. 2013. CN Patent. 103215662A.
[23] Zhang YC, Li XH, Zhong XF, et al. High-speed preparation device and process for centrifugal electrospinning nano-twisting yarn. 2013. CN103409861A.
[24] Mindru TB, et al. Morphological aspects of polymer fiber mats obtained by airflow rotary-jet spinning. Fibers and Polymers 2013;14(9):1526–34.
[25] Zou SB, Yuan J. Centrifugal electrostatic continuous spinning nanofiber device. 2014. CN104178826A.
[26] Liu Y, Li XH, He H, et al. A new type of centrifugal melt electrospinning device. 2014. CN104088024A.

[27] Peng H. Experiment and Simulation of centrifugal melt electrospinning, Beijing University of Chemical Technology, 2017.

[28] Jin CK, Zhong QR. Application of Fourier transform in nanofiber orientation measurement. Journal of Textile Research 2013;34(11):34–8.

[29] Liu SL. Assembly of oriented ultrafine polymer fibers by centrifugal electrospinning. Journal of Nanomaterials 2013:1–9.

[30] Edmondson D, Cooper A, Jana S, et al. Centrifugal electrospinning of highly aligned polymer nanofibers over a large area. Journal of Materials Chemistry 2012;22(35):18646.

[31] Zander NE. Formation of melt and solution spun polycaprolactone fibers by centrifugal spinning. Journal of Applied Polymer Science 2015;132(2):41269.

[32] Erickson AE, Edmondson D, Chang FC, et al. High-throughput and high-yield fabrication of uniaxially-aligned chitosan-based nanofibers by centrifugal electrospinning. Carbohydrate Polymers 2015;134:467–74.

[33] Ayres C, Bowlin GL, Henderson SC, et al. Modulation of anisotropy in electrospun tissue-engineering scaffolds: analysis of fiber alignment by the fast Fourier transform. Biomaterials 2006;27(32):5524–34.

[34] Khamforoush M, Asgari T, Hatami T, et al. The influences of collector diameter, spinneret rotational speed, voltage, and polymer concentration on the degree of nanofibers alignment generated by electrocentrifugal spinning method: modeling and optimization by response surface methodology. Korean Journal of Chemical Engineering 2014;31(9):1695–706.

[35] Liu Y, Peng H, Pan K, et al. Combined wired and wireless temperature messurement type centrifugal melt electrostatic spinning device 2017. CN104088024A.

Chapter 7

Dissipative particle dynamics simulations of centrifugal melt electrospinning

Chapter outline

7.1 Introduction

Centrifugal electrospinning is a combination of centrifugal spinning and high-voltage electrospinning. When the spinning starts, the solution or melt is ejected from the nozzle by the centrifugal force, following which the resulting jet moves continuously with the electric field force and gravity [1]. Compared with traditional centrifugal spinning, centrifugal electrospinning requires a lower rotation speed, which avoids the requirement for a high-speed motor. Moreover, compared with normal electrospinning, a high voltage is not required. Generally, centrifugal electrospinning is not only conducive for the preparation of ordered or aligned nanofibers but also improves the spinning efficiency and allows the achieving of mass production. However, irrespective of the type of electrospinning until a solvent (except water) is used, it is impossible to avoid the environmental pollution caused by solvent

Melt Electrospinning. https://doi.org/10.1016/B978-0-12-816220-0.00007-5

evaporation, high cost, difficult solvent selection, and toxicity of the produced fiber. Melt electrospinning was invented by Larrondo and Manley [2], and many other researchers have also made tremendous progress [3–5]. For melt electrospinning, it is difficult to produce fibers on the nanometer scale, which has become a bottleneck in its further development and industrial application. According to our preliminary research [1], the diameter of a fiber obtained by centrifugal melt electrospinning is much smaller than that synthesized by ordinary melt electrospinning (up to several hundred nanometers), and the production efficiency of the former method is significantly improved. In addition, industrialization is less difficult for the combined method than for the common centrifugal spinning or electrospinning. Therefore, centrifugal melt electrospinning is a highly feasible method for solving the above bottleneck problem. To study the spinning process rapidly and efficiently, we explored the evaluation trends of the fiber diameter, yield, and molecular chain length in electrostatic fields or pulsed electric fields by dissipative particle dynamics (DPD) simulation.

DPD is a mesoscale numerical simulation method developed and based on molecular dynamics and gas lattice, which was originally proposed by Hoogerbrugge and Koelman [6–8] and subsequently improved by Español and Warren [9–11], who laid the foundation for the DPD simulation method. The DPD method has made significant progress in simulating real systems, which has significantly advanced its various applications in simulation methods. The relationship between the simulation and interaction parameters in Flory–Huggins theory is calculated. To accelerate the calculation, DPD adopts a coarse-grained model to eliminate the structural information of the microscopic molecules. DPD has excellent performance in many complex fields, including migration behavior of deformable droplets [12], morphology of hydrated Nafion [13], polymer polar cells [14], phase behavior [15], electrospinning [16], etc. In this study, the centrifugal melt electrospinning process was simulated using the DPD method.

7.2 The dissipative particle dynamics model in centrifugal melt electrospinning

In this research, the model system is developed based on our previous studies [17–20]. For the centrifugal electrospinning process, the charge is simulated by selectively setting charged particles on the nodes of the different polymer chains in the simulation system. Subsequently, the effect on the fiber velocity and trajectory is studied by calculating the change in the number of charges from one to n (n is the number of particles in the entire chain). Simultaneously, the charged particles can be set at the end of the initial stage of the jet to simulate the force at its one end.

In the simulation system shown in Fig. 7.1C, the red particles are under four types of mesoscopic forces: (1) electric field force $\vec{F}_e$, which represents the

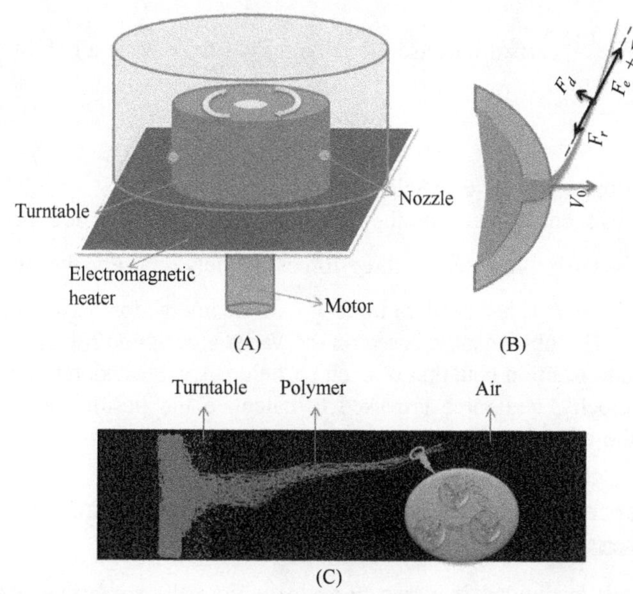

FIGURE 7.1 (A) Schematic diagram of centrifugal melt electrospinning device and DPD model centrifugal electrospinning device, (B) forces of fiber, (C) model in DPD[21].

force of the charged particles in the electric field; (2) centrifugal force $\vec{F}_c$, which is the inertial force of the particle due to rotation; (3) Coriolis force $\vec{F}_d$, which reflects the inertia of particle motion; and (4) resistance $\vec{F}_r$, which emulates the friction between particles. All the forces follow Newton's motion equation:

$$m d\vec{v}/dt = \vec{F}_e - \vec{F}_r + \vec{F}_c - \vec{F}_d \tag{7.1}$$

where, $m = 1$, $\vec{v}$ is the absolute speed of the droplet, and its direction changes with the particle motion, and t is the motion time.

Electric field force $\vec{F}_e$ is related to field strength $\vec{E}$ and charge q, i.e.,

$$\vec{F}_e = \vec{E} q \vec{r}/|\vec{r}| \tag{7.2}$$

where, $\frac{\vec{r}}{|\vec{r}|}$ represents a unit vector of the electric field.

Centrifugal force $\vec{F}_c$ causes the jet to move in a more distant direction, which is related to the angular velocity $\vec{\Omega}$:

$$\vec{F}_c = 2m\vec{\Omega} \times \vec{v} \tag{7.3}$$

where, m is the particle mass, assuming $m = 1$.

Coriolis force $\vec{F}_d$ refers to the offset of the linear motion of an object relative to the rotating system. The formula is as follows:

$$\vec{F}_e = 2m\vec{\Omega} \times \vec{\Omega} \times \vec{r} \tag{7.4}$$

where, $\vec{r}$ indicates the position of the jet particles and $m = 1$.

Resistance $\overrightarrow{F}_r$ arises because the polymer is subjected to air and there is a viscous resistance:

$$\overrightarrow{F}_r = \eta(\overrightarrow{v} - \overrightarrow{\Omega} \times \overrightarrow{r}) \tag{7.5}$$

where, η is the viscosity coefficient.

To achieve equilibrium in the system, each coarse-grained particle in the DPD model is subjected to three forces, namely, conservative force $\overrightarrow{F}^C$, dissipative force $\overrightarrow{F}^D$, and random force $\overrightarrow{F}^R$ of the interaction between particles i and j [22–25]. Subsequently, based on the Varlet algorithm [26] and Eq. (7.1), the coordinate position equation of each particle is established, and the current position, velocity, and force are used to calculate the position and velocity change at the next moment.

7.3 Different electric field simulation of centrifugal melt electrospinning

This research is divided into two parts to explore the performance of centrifugal melt spinning fiber. On the one hand, it studies an ordinary electrostatic field, on the other, it examines the variation law of a jet under a pulsed electric field. Fig. 7.2 is a waveform diagram of a common electric field and a pulsed electric field. Under the ordinary electric field, the jet particles move toward the direction of the electric field force (i.e., the polymer molecules move under the field strength), and the electric field force is linear so that the jet has a stretching effect. However, under the pulsed electric field, it can be seen from Fig. 7.2 that the electric field is constantly changing. In one cycle, there are two states of a maximum positive charge and no charge on the jet.

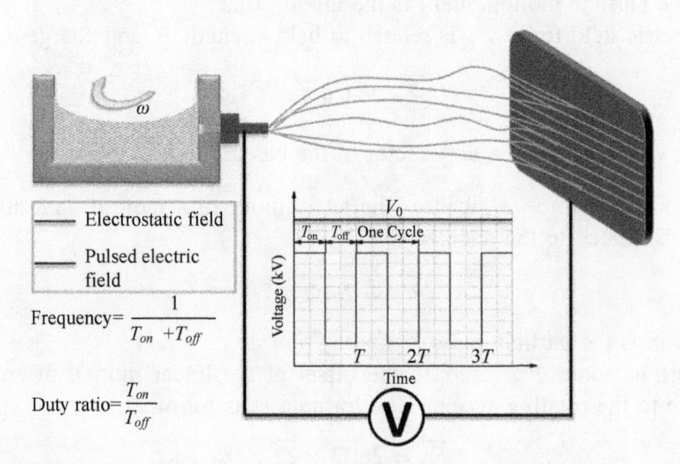

FIGURE 7.2 Schematic diagram of two types of electric fields and the definitions of frequency and duty ratio [21].

This causes the molecular chain to be more easily unwound under the action of the repulsion between the molecular chains and axial stretching and relaxation. Therefore, in the second part, we use two characteristic parameters (duty cycle and frequency) of the pulsed electric field for analyzing its effect on the jet.

7.3.1 Centrifugal melt electrospinning in an electrostatic field

In the centrifugal melt electrospinning process, a fiber is subjected to complex forces such as centrifugal force, electrostatic force, gravity, air resistance force, surface tension force, and viscoelastic force. These forces are in varied planes, and their directions are also different. The various dominant positions of the different forces will form a variety of fiber path trajectories, and the most important tensile powers should be the centrifugal force and the electrostatic force. The fiber jet moves in a circular area between the turntable and the receiving ring. The centrifugal and electrostatic forces cause the fibers to move in a circular motion and radially, respectively. A game behavior occurs between the centrifugal force and electrostatic force, and the combined force produces a corresponding fiber trajectory. In addition, the temperature directly affects the molten state of the polymer particles, thereby becoming a prerequisite for the electrospinning of the centrifugal melt. Various theoretical problems in centrifugal melt electrospinning, such as molecular chain motion, are best investigated by DPD simulations. In this study, the jet diameter, productivity, and molecular chain length are qualitatively analyzed based on three important parameters, system temperature, centrifugal speed, and electrostatic coefficient.

7.3.1.1 Effect of electrospinning parameters on jets

In experiments, fiber morphology can be observed by scanning electron microscopy (SEM), and subsequently, fiber diameter statistics and analysis can be performed using software Image J and Origin 8.0 [27–29]. However, in the simulation system, the jet is composed of virtual dotted lines and a single jet contains multiple molecular chains, so it is not suitable to measure directly, using Image J. Therefore, for the simulation system, an appropriate method is adopted to define the jet diameter. In Fig. 7.2A, D is the average diameter of the jet, $\sqrt{a^2+b^2}$ is the calculated length of the jet segment, and N is the total number of particles in the jet segment. Thus, the magnitude of the jet diameter is represented by the calculated value of the average density of the particles in the jet segment. The jet diameter was measured in the last stage of the moving jet, which was exactly above the collector or the bottom of the simulation system. The average diameter was calculated from the same stage of 20 repeated simulations of one case. The simulations were performed repeatedly to obtain a statistical result because of the DPD simulation including a random force [30].

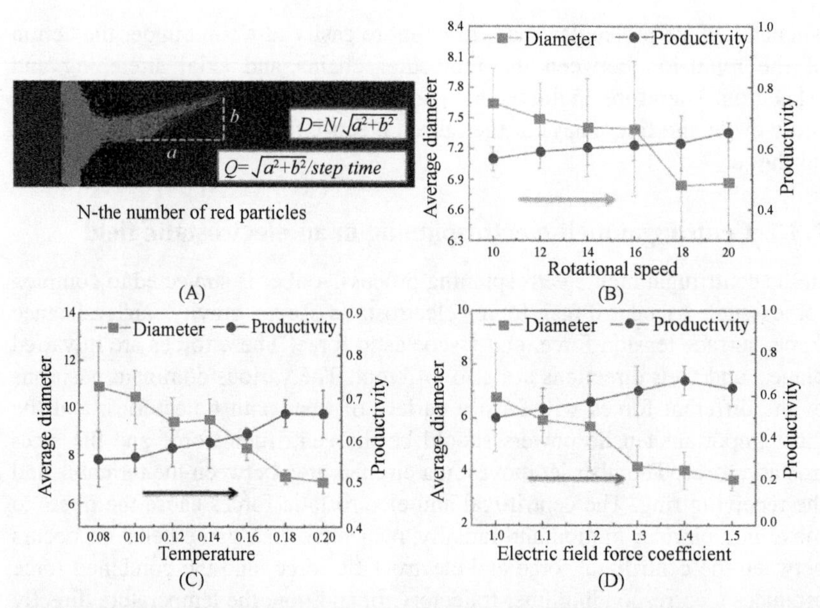

FIGURE 7.3 (A) Schematic diagram of jet diameter calculation; and changes of average diameter and productivity with three factors, (B) rotational speed, (C) temperature, (D) electric field force coefficient [21].

For electrospinning, the productivity of nanofibers is important, particularly for achieving a yield on a large scale. In the simulation, the productivity was defined as the number of fibers produced per unit time. Fiber productivity (Q) is calculated as the drop distance of the fibers at the same time step of different flying processes. As shown in Fig. 7.3A, $Q = \sqrt{a^2 + b^2}/\text{step time}$, where step time is the same, in both cases to calculate productivity, the 559th step time is chosen. A longer drop distance at the same time step leads to faster fiber movement and higher fiber yield.

As can be seen from Fig. 7.3B, with the increase in rotational speed, the average diameter of the jet firstly decreases and then increases. Consequently, under the tension of the centrifugal force, the jet reacts with the air resistance and its viscous force (conservative force). Thus, it undergoes stretching under the multiple internal and external forces, and so the jet diameter reduces with an increase in the centrifugal force. However, when the rotation speed is very high, some red particles can reach the receiving plate without stretching. Thus, the diameter of the jet has a relatively large variation.

In centrifugal melt electrospinning, in addition to the rotational speed, the temperature is a key factor affecting the morphology and performance of the fibers. The change in the jet diameter was studied at different system temperatures, as shown in Fig. 7.3C. It can be seen from the figure that with the increase in temperature, the jet diameter gradually decreases. Thus, from the perspective of the simulation system, the dissipation force of the system reduces with the temperature rise. From an experimental viewpoint, as the temperature of the system increases, the viscosity of the polymer decreases, increasing the activity of the red particle movement more actively and causing the fibers to be more easily stretch.

As shown in Fig. 7.3D, the increase in the electric field force coefficient causes the fiber diameter to gradually increase. Because the electric field coefficient reflects the magnitude of the electric field force, a larger electric field force implies more stretching of the jet, so that the average diameter of the jet becomes reduced. However, in our experience, when the voltage is sufficiently high, air breakdown is likely to occur. Unfortunately, this phenomenon cannot be reflected by the DPD simulation, which also illustrates its limitations.

It can be seen from the trend of the blue line graph in Fig. 7.3B–D that the fiber productivity is increasing as the rotational speed, temperature, and electric field force coefficient are gradually increasing. Among these, when the temperature coefficient and dielectric constant are 0.18 and 1.4, respectively, the fiber productivity has an inflection point, showing a decreasing trend. This trend indicates that the electric field force and temperature increase can actually enhance the fiber yield, but if the value hypertrophies, then the polymer melt is limited by the small nozzle. More melt will increase the pressure on the nozzle wall and thereby reduce the disentanglement. This is our hypothesis for the productivity decrease. In the future, we will perform some experiments to verify the above explanation.

7.3.1.2 Effect of electrospinning parameters on molecular chain untangling

Entanglement is one of the important characteristics of a polymer. The complex hierarchical structure and metastable nature of polymers have a major relationship with the entanglement network [31–34]. The entanglement is related to the crystallization, rheological properties, and glass transition of the polymer. To observe molecular chain untangling, in the centrifugal melt electrospinning simulation model, some molecular chains were chosen. Their movement trajectories were traced and analyzed. Molecular chain stretching or chain length is an important physical factor that can reflect the internal structure of a jet. Generally, the entangled chains have a short length, whereas untangled chains are long. Therefore, we select the chain length as a qualitative measure of the entanglement occurring in this simulation. The DPD method records the detailed evaluation information of each particle. Therefore, it allows the measurement of the chain length during any period in the process. The mean square terminal distance of a single molecular chain is taken as the chain length. For quantitative analysis, the length of the polymer chains in the centrifugal electrospinning process

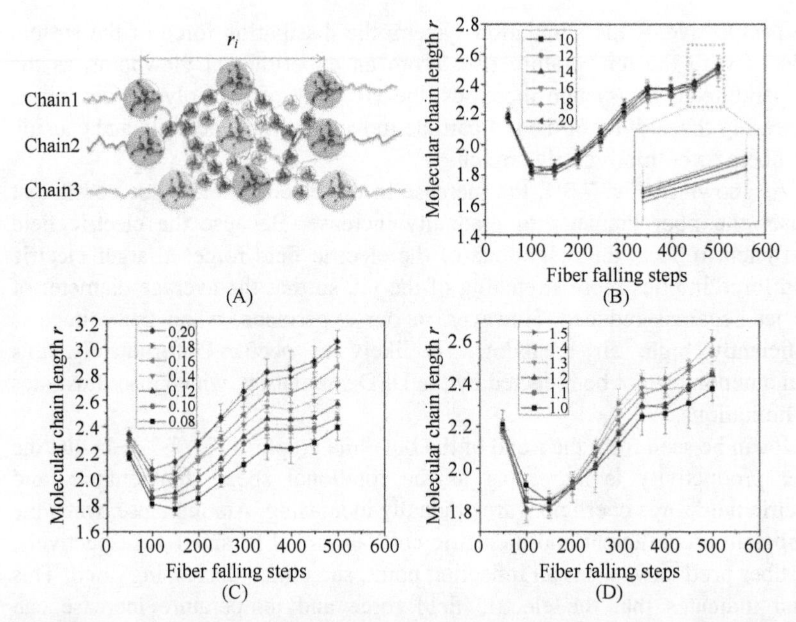

FIGURE 7.4 (A) Schematic of molecular chain entanglement and changes of molecular chain length at different, (B) rotational speeds, (C) temperature coefficients, (D) electric field force coefficients [21].

was calculated at the same time step (Fig. 7.4A). To reduce the error, the mean square end-to-end distances of 20 molecular chains, at the same time step, were selected and averaged. The calculation results are integrated into a line chart as shown in Fig. 7.4B–D. Accordingly, the mean square end-to-end distance is given by:

$$r = \sqrt{\left(r_1^2 + r_2^2 + r_3^2 + \cdots + r_i^2\right)/n} \tag{7.6}$$

It can be obtained from the curves in Fig. 7.4B–D, when the temperature and electric field force increase, that the molecular chain length increases. A higher temperature reduces the viscous resistance of the melt, allowing the polymer chains to more easily stretch into fibers. When the electric field force becomes great, the amplitude of the jet whipping reduces. The increased force leads to easier unwinding of the molecular chain. However, a maximum tensile force is observed at a rotational speed of 18, as shown in the inset of Fig. 7.4B. Raising the rotational speed implies an increase in the centrifugal force of the jet, which causes stretching of the fibers. Consequently, the molecular chain length of the jet becomes longer as the rotational speed increases. However, when the rotational speed is extremely high, the jet is not sufficiently stretched per unit time because the speed is very fast, and so, the fiber diameter becomes smaller. Comparing the above three figures, it can also be seen that when the step is 50–100, the molecular chain length first decreases, and then increases. The reason for this phenomenon can be explained by the fact that in the beginning, the polymer collects at the nozzle or in the process of the forming of the Taylor cone. Subsequently, the polymer aggregates and the chain lengths are reduced. After it passes through the nozzle, the chain is stretched by the forces and the length starts to increase. In addition, the curves show that in the present simulation system, the difference in the chain lengths at varied rotational speeds and electric field forces is not obvious, whereas the length at different temperatures is significantly different. This indicates that the temperature plays an important role in the disentanglement of the chains.

7.3.2 Centrifugal melt electrospinning in a pulsed electric field

The most common form of the electric field in electrospinning is a constant form that does not change. In recent years, pulsed electric fields have shown distinct advantages in some aspects [35,36]. A pulsed electric field can be understood as an electric field whose voltage changes in its waveform. Because a melt has low conductivity and high viscosity, the diameter of the spun fiber is relatively coarse. Research has shown that a pulsed electric field can effectively reduce the diameter of electrospinning fiber [35,36]. However, in a pulsed electric field, the jet is not electrically neutral, having the same type of electric charge at each part along the fiber. Therefore, there is a repulsive force between all the adjacent fiber parts. In a pulsed electric field, under the action of the interfiber part repulsion and axial relaxation and stretching, a molecular chain is more easily disentangled. Thus, it is predicted that such a field promotes the refinement of melt electrospun fibers. However, a pulsed electric field has similar charges as an alternating electric field (AC), except that the net charge of the jet in the latter is zero so that the charges have a completely different effect on the jet [37–39]. When a pulsed electric field is applied, the spinning conditions change. This study focuses on the effects of the frequency and duty cycle of a pulsed electric field on the molecular chain stretching and jet diameter.

7.3.2.1 Effects of duty ratio on jet

In a pulsed electric field, the duty ratio is a part of the period when the electric field force is applied, as shown in Fig. 7.2. A larger duty ratio implies a longer operation time of the electric field in the same period. Similar to the above simulation, the average diameter was calculated from the same stage of 20 repeated simulations of one case.

Fig. 7.5A presents the morphology of the fiber dropping in the 559th step when the duty ratio is 50%, 60%, 70%, 80%, 90%, and 100% in a

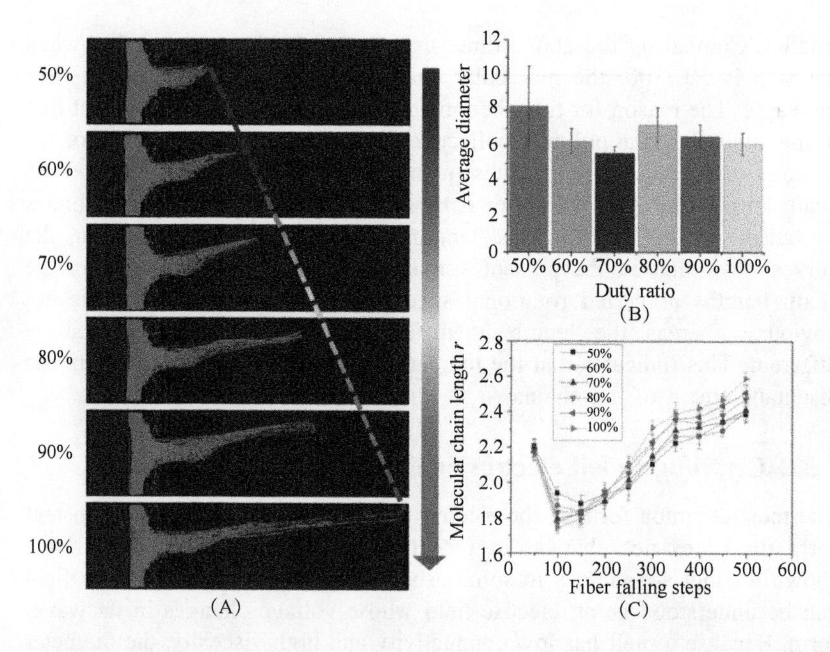

FIGURE 7.5 Effect of duty ratio on a spun jet: (A) the morphology of fiber dropping; and change of (B) average diameter with duty ratio and (C) molecular chain length with falling steps [21].

pulsed electric field. As can be seen from Fig. 7.5A, increasing the duty ratio accelerates the dropping velocity of the fiber. However, as shown in Fig. 7.5B, as the duty ratio increases, the jet diameter first decreases and then increases. The jet diameter is small when the duty ratio is 70%. A possible reason is that in the jet flowing process, although a long time of activity of the electric field force is advantageous for fiber refining when the duty ratio is larger than a certain value, the number of the polymers that are simultaneously stretched at the spout nozzle is increased drastically. In the same cycle, the action of the long-term electric field force has a dual effect on the jet, and the resulting fiber diameter becomes larger instead. In the present simulation, the duty ratio of 70% is more appropriate for the polymer.

The changes in the molecular chain length during the centrifugal melt electrospinning are shown in Fig. 7.5C. The elongation of the molecular chain is represented by the mean square end-to-end distance r. The curves show that the duty ratio has a significant effect on the molecular chain length. At duty ratio 100%, the degree of stretching is large. This is consistent with Fig. 7.5A. This phenomenon can be explained in that a larger duty ratio can provide a longer action time in the pulsed electric field, and that the electric field force can overcome the resistance so that the molecular chain has sufficient time to stretch.

7.3.2.2 Effects of frequency on jet

The movement of a molecular chain in a pulsed electric field is a complex disentanglement process, and because this motion is microscopic, it is actually difficult to observe. Therefore, we visually observe the molecular chain motion via the simulation. As shown in Fig. 7.6A, the jet diameter decreases with increasing frequency. On the one hand, at the same intensity and duty ratio of the electric field, a high frequency allows the pulsed voltage to act rapidly on the jet change, causing the jet to accumulate more energy. This increases the tensile kinetic energy, allowing the polymer melt chains to untangle easily and ultimately cause the average diameter of the jet to reduce.

On the other hand, as seen in Fig. 7.6B, when the frequency of the electric field increases, the molecular chain length increases. From frequency 3.0 to 5.0, the length increases sharply. This indicates that the frequency increase is in favor of the untangling of the molecular chains.

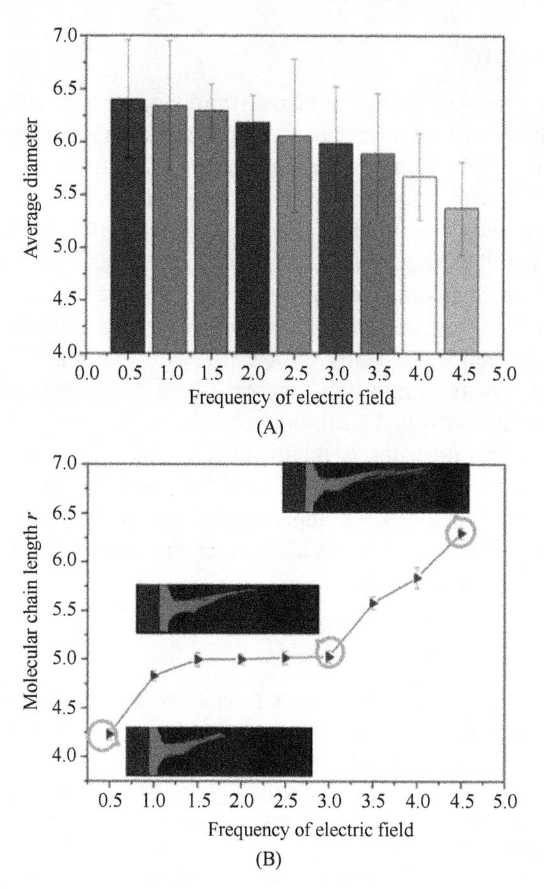

FIGURE 7.6 Changes of (A) average diameter and (B) molecular chain length with frequency [21].

This phenomenon can be explained by the fact that in a pulse period, i.e., half a cycle of the voltage, the molecular chain in the jet is stretched under the action of the electric field force. In the half cycle without the voltage action, the jet is decelerated owing to the force reduction, and the molecular chain is relaxed. This promotes the entanglement of the molecular chains. Compared with Fig. 7.4, the molecular chain length at different frequencies of a pulsed electric field is longer than that in a normal electric field. This proves that a pulsed electric field can promote melt electrospinning jet stretching. This is because the molecular chains in a pulsed electric field have the same type of charges. Hence, adjacent chains and segments repel. The change forces between the molecular chains allow the disentanglement to occur rapidly. Therefore, we achieve a finer jet. This is also consistent with the results of Fig. 7.6A, i.e., an increase in the molecular chain length results in a smaller jet diameter.

7.4 Conclusion

In this study, a pulsed electric field was introduced in centrifugal melt electrospinning. Its effects on the molecular chain stretching and jet diameter were explored by the DPD simulation method. By trial and error, the appropriate centrifugal melt electrospinning model was established. Jets in common and pulsed electric fields were studied. The results show that in the electrostatic field, increasing the rotational speed, temperature, and electric field force could reduce the fiber diameter and increase the fiber yield. Simultaneously increasing the temperature and electric field force, the molecular chain of the fiber were lengthened gradually. However, as the rotational speed increased, the molecular chain length first increase and then decrease. This also illustrates that in centrifugal melt electrospinning, the effect of the rotational speed on the fiber properties is relatively complex. In a pulsed electric field, when the duty ratio is 70%, the fiber diameter is the shortest; when the duty ratio is 100%, the fiber molecular chain is the longest. With the increase in the frequency, the average fiber diameter and molecular chain length exhibits opposite trends.

References

[1] Wu S, Peng H, Li X, Streubel PN, Liu Y, Duan B. Effect of scaffold morphology and cell co-culture on tenogenic differentiation of HADMSC on centrifugal melt electrospun poly (L-lactic acid) fibrous meshes. Biofabrication 2017;9(4):0441064.

[2] Larrondo L, John Manley RS. Electrostatic fiber spinning from polymer melts. I. Experimental observations on fiber formation and properties. Journal of Polymer Science: Polymer Physics Edition 1981;6(19):909–20.

[3] Visser J, Melchels FPW, Jeon JE, et al. Reinforcement of hydrogels using three-dimensionally printed microfibres. Nature Communications 2015;6:6933.

[4] Castilho M, Hochleitner G, Wilson W, et al. Mechanical behavior of a soft hydrogel reinforced with three-dimensional printed microfibre scaffolds. Scientific Reports 2018;8:1245.

[5] Martine LC, Holzapfel BM, McGovern JA, et al. Engineering a humanized bone organ model in mice to study bone metastases. Nature Protocols 2017;12(4):639–63.

[6] Li T, Jiang Z, Yan D, Nies E. A polyethylene chain investigated with replica exchange molecular dynamics simulation: equilibrium lamellar thickness and melting point, ordering and free energy. Polymer 2010;51(23):5612–22.

[7] Chudoba R, Heyda J, Dzubiella J. Temperature-dependent implicit-solvent model of polyethylene glycol in aqueous solution. Journal of Chemical Theory and Computation 2017;13(12):6317–27.

[8] Español P, Warren PB. Perspective: dissipative particle dynamics. The Journal of Chemical Physics 2017;146(15):1–166.

[9] Groot RD, Warren PB. Dissipative particle dynamics – bridging the gap between atomistic and mesoscopic simulation. The Journal of Chemical Physics 1997;11(107): 4423–35.

[10] Zhou B, Luo W, Yang J, et al. Simulation of dispersion and alignment of carbon nanotubes in polymer flow using dissipative particle dynamics. Computational Materials Science 2017;126:35–42.

[11] Hu J, Zhang C, Li X, Han J, Ji F. Architectural evolution of phase domains in shape memory polyurethanes by dissipative particle dynamics simulations. Polymer Chemistry 2017;8(1):260–71.

[12] Marson RL, Huang Y, Huang M, Fu T, Larson RG. Inertio-capillary cross-streamline drift of droplets in Poiseuille flow using dissipative particle dynamics simulations. Soft Matter 2018;14(12):2267–80.

[13] Liu H, Cavaliere S, Jones DJ, Roziere J, Paddison SJ. Morphology of hydrated nafion through a quantitative cluster Analysis: a case study based on dissipative particle dynamics simulations. Journal of Physical Chemistry C 2018;122(24):13130–9.

[14] Du C, Ji Y, Xue J, et al. Morphology and performance of polymer solar cell characterized by DPD simulation and graph theory. Scientific Reports 2015;5:16854.

[15] Lu T, Guo H. Phase behavior of lipid bilayers: a dissipative particle dynamics simulation study. Advanced Theory and Simulations 2018;5(1):1–13.

[16] Liu Y, Wang X, Yan H, Guan C, Yang W. Dissipative particle dynamics simulation on the fiber dropping process of melt electrospinning. Journal of Materials Science 2011;46(24):7877–82.

[17] Song Q, Zhang J, Liu Y. Mesoscale simulation of a melt electrospinning jet in a periodically changing electric field. Chemical Journal of Chinese Universities 2017;38(6):966–74.

[18] Wang X, Liu Y, Zhang C, An Y, He X, Yang W. Simulation on electrical field distribution and fiber falls in melt electrospinning. Journal of Nanoscience and Nanotechnology 2013;13(7):4680–5.

[19] Liu Y, Kong B, Yang X. Studies on some factors influencing the interfacial tension measurement of polymers. Polymer 2005;46(8):2811–6.

[20] Liu Y, Yang X, Yang M, Li T. Mesoscale simulation on the shape evolution of polymer drop and initial geometry influence. Polymer 2004;45(20):6985–91.

[21] Li K, Xu Y, Liu Y, Mohideen MM, He H, Ramakrishna S. Dissipative particle dynamics simulations of centrifugal melt electrospinning. Journal of Materials Science 2019; 54(13): 9958–68.

[22] Lísal M, Šindelka K, Suchá L, Limpouchová Z, Procházka K. Dissipative particle dynamics simulations of polyelectrolyte self-assemblies. Methods with explicit electrostatics. Polymer Science 2017;59(1):77–101.

[23] Alizadehrad D, Fedosov DA. Static and dynamic properties of smoothed dissipative particle dynamics. Journal of Computational Physics 2018;356:303–18.

[24] Wang Z, Quik JTK, Song L, Wouterse M, Peijnenburg WJGM. Dissipative particle dynamic simulation and experimental assessment of the impacts of humic substances on aqueous aggregation and dispersion of engineered nanoparticles. Environmental Toxicology and Chemistry 2018;37(4):1024–31.

[25] Ketkaew R, Tantirungrotechai Y. Dissipative particle dynamics study of SWCNT reinforced natural rubber composite system: an important role of self-avoiding model on mechanical properties. Macromolecular Theory and Simulations 2018;27(3):1700093.

[26] Plimpton S. Fast parallel algorithms for short-range molecular dynamics. Journal of Computational Physics 1995;117(1):1–19.

[27] Liu SL. Assembly of oriented ultrafine polymer fibers by centrifugal electrospinning. Journal of Nanomaterials 2013:8 (2514103).

[28] Erickson AE, Edmondson D, Chang F, et al. High-throughput and high-yield fabrication of uniaxially-aligned chitosan-based nanofibers by centrifugal electrospinning. Carbohydrate Polymers 2015;134:467–74.

[29] Liu S, Sun B, Yin H, et al. Fabrication of fluorescent polymer crossbar arrays and micro-ropes via centrifugal electrospinning. In: Yu L, Guo WP, Sun M, He J, editors. Advanced Materials Research, vol. 517; 2013. p. 785–6.

[30] Li Z, Yuan Y, Chen B, et al. Photo and thermal cured silicon-containing diethynylbenzene fibers via melt electrospinning with enhanced thermal stability. Journal of Polymer Science Part A Polymer Chemistry 2017;55(17):2815–23.

[31] Luo C, Kröger M, Sommer J. Molecular dynamics simulations of polymer crystallization under confinement: entanglement effect. Polymer 2017;109:71–84.

[32] Hagita K, Morita H, Takano H. Molecular dynamics simulation study of a fracture of filler-filled polymer nanocomposites. Polymer 2016;99:368–75.

[33] Sliozberg YR, Mrozek RA, Schieber JD, et al. Effect of polymer solvent on the mechanical properties of entangled polymer gels: coarse-grained molecular simulation. Polymer 2013;54(10):2555–64.

[34] Kröger M. Shortest multiple disconnected path for the analysis of entanglements in two- and three-dimensional polymeric systems. Computer Physics Communications 2005;168(3):209.

[35] Xie G, Wang Y, Han X, et al. Pulsed electric fields on poly-l-(lactic acid) melt electrospun fibers. Industrial and Engineering Chemistry Research 2016;55(26):7116.

[36] Li K, Wang Y, Xie G, et al. Solution electrospinning with a pulsed electric field. Journal of Applied Polymer Science 2018;135(15):46130.

[37] Xie J, Jiang J, Davoodi P, Srinivasan MP, Wang C. Electrohydrodynamic atomization: a two-decade effort to produce and process micro-/nanoparticulate materials. Chemical Engineering and Science 2015;125:32–57.

[38] Sarkar S, Deevi S, Tepper G. Biased AC electrospinning of aligned polymer nanofibers. Macromolecular Rapid Communications 2010;28(9):1034–9.

[39] Pokorny P, Kostakova E, Sanetrnik F, et al. Effective AC needleless and collectorless electrospinning for yarn production. Physical Chemistry Chemical Physics 2014;16(48):26816–22.

Chapter 8

Three-dimensional (3D) printing based on controlled melt electrospinning in polymeric biomedical materials

Chapter outline

8.1 Introduction

Biomaterial designs attempt to harness the regenerative capacity of the body, by merging principles of materials engineering and biological science, to repair damaged tissue in the body [1–3]. Over the last few decades, several multimodal biomimetic strategies have emerged to alleviate damaged tissue using a wide range of fabrication methods [4,5]. Of these, electrospinning has gained rapid recognition, for opening a new horizon in tissue engineering methodologies, owing to its simple yet precise methods to fabricate scaffolds with nano/macroscale topography [6–9]. Near-field electrospinning is a relatively new technique that utilizes an electrically charged polymer solution to deposit continuous fibrous meshes [10–12]. These meshes are porous, biocompatible, have a high surface area, and can be fabricated with advanced features such as drug elution of orientation from a range of polymers.

Melt Electrospinning. https://doi.org/10.1016/B978-0-12-816220-0.00008-7

Such scaffolds ultimately resemble the structure and size of the extracellular matrix natural tissues. This has made electrospinning an attractive strategy to produce surgical constructs for regenerative medicine. Although near-field electrospinning, as a concept, was proposed in 2006 [13], direct-write electrospinning research has experienced exponential growth in the past decade. This is attributed to the accuracy of the printing technique in recapitulating the micro/nanoscale composite structure required to meet the needs of individual patients [14–19].

In recent years, scientists have combined three-dimensional (3D) printing and electrospinning, and achieved significant milestones in biomedical design methods for tissue engineering. Many multidisciplinary research teams across the world have advanced the electrospun 3D printed precision and controllability, porosity, and mechanical properties of the scaffolds. Unlike nonbiological applications, 3D bioprinting with electrospinning involves posing significant technical challenges to cater for the sensitivity of living cells and the construction of functional tissues. Fabrication of such designs requires high geometric accuracy as well as the incorporation of complexities such as biomaterial choice, cell types, and bioactivity factors [20,21].

Three-dimensional printing in biomedical polymeric materials can be broadly split into four basic levels; (1) organic model manufacturing, (2) permanent implacable manufacturing, (3) indirect assembly of cells, and (4) direct cell assembly manufacturing [22–26]. At present, 3D printing has already found application in surgical analysis planning and manufacturing of prosthetic implants [27–31]. The augmentation of scaffold fabrication with 3D printing stands to deliver enormous sophistication and personalized solutions to tissue damage. Due to the high precision of the technology [15,32,33], electrospinning combined with 3D printing continues to evolve, in the hope of achieving specific outcomes in regenerative medicine through scaffold engineering, drug delivery, wound dressing, and enzyme immobilization [34–41]. With the rise in the aging population and increase in regenerative medical demands, the development of scalable and automated bioprinters will enormously impact the quality and affordability of medical care in the future.

In this review, we provide an in-depth understanding of 3D printer devices, with a focus on electrospinning underlining techniques, to produce fibers for biological morphology and application. Herein, we discuss the basic working principles and compare several additional features, such as electrostatic lens auxiliary electrodes, core–shells, the core structure of spray nozzles, and needle core induction receivers.

8.2 Basic principles of 3D printing based on electrospinning

Three-dimensional printing based on electrospinning is a fabrication device that combines near-field electrospinning with a computer-aided design (CAD)/computer-aided manufacturing (CAM) system. Near-field electrospinning drives the formation of fibers through the stretching of melted or dissolved polymer into the Taylor cone. The polymer melt or solution is subjected to a high-voltage electrostatic field and the excited micro/nano jet produces direct-writing of a single fiber of the needle-collector at an appropriate distance. In 3D printing with electrospinning, the single fiber produced is particularly superimposed on the dielectric plate collectors geometrically assisted by the CAD/CAM component. As the polymer solution is electrostatically drawn into a jet following an initially vertical stability zone, bending instabilities develop in a second zone along the flight path of the charged jet [14,42]. However, the range of the needle-collector distance in the direct-writing method, although in an initially straight stable zone, is limited.

There are considerable differences in the setup of melt electrospinning writing (MEW) and solution electrospinning (SES). Fig. 8.1 represents a schematic overview of a 3D printing device with solution and melt electrospinning, which is widely used for manufacturing support [44] and tissue engineering [45]. Melt electrospinning writing is largely based on fused deposition modeling (FDM) combined with near-field electrostatics that ride the diameter of the fiber. The setup uses a heating coil along with temperature control in order to provide enough heat to melt the polymer. The raw material commonly used in melt electrospinning writing is granular material, while filamentous thermoplastic material is usually used for FDM technology [46], as shown in Fig. 8.1C [43]. Recently, researchers have developed a granular material that is especially designed for 3D printers. Melt electrospinning writing has a higher viscosity and lower electrical conductivity, requiring the needle-collector distance to be of a wider range [47]. Melt electrospinning writing bypasses medical translational challenges such as solvent toxicity and accumulation associated with SES systems [48]. While both methods provide orderly structures, charge accumulation in SES impacts the number of layers causing the fibers to be one coherent structure. This greatly impacts the average pore size of the scaffold and overall quality of a tissue engineered product. On the other hand, melt electrospinning writing allows a larger pore size of the scaffold produced that is required for optimal cell invasion and growth. So far, melt electrospinning writing has been used to create three-dimensional bone scaffold [49], vascular grafts [50], and bone cell implants [51].

8.3 Different auxiliary electrode and dielectric plate collectors

The traditional electrospinning process is fairly unstable and almost uncontrollable due to the bending of a charged jet under coupled multifield forces. A 3D printing device with near-field electrospinning helps to overcome this hurdle and improve the fine control over the construction design. In addition, to narrow down the collector distance, it is now possible to stabilize the jet by increasing the auxiliary electrode and changing the collector device.

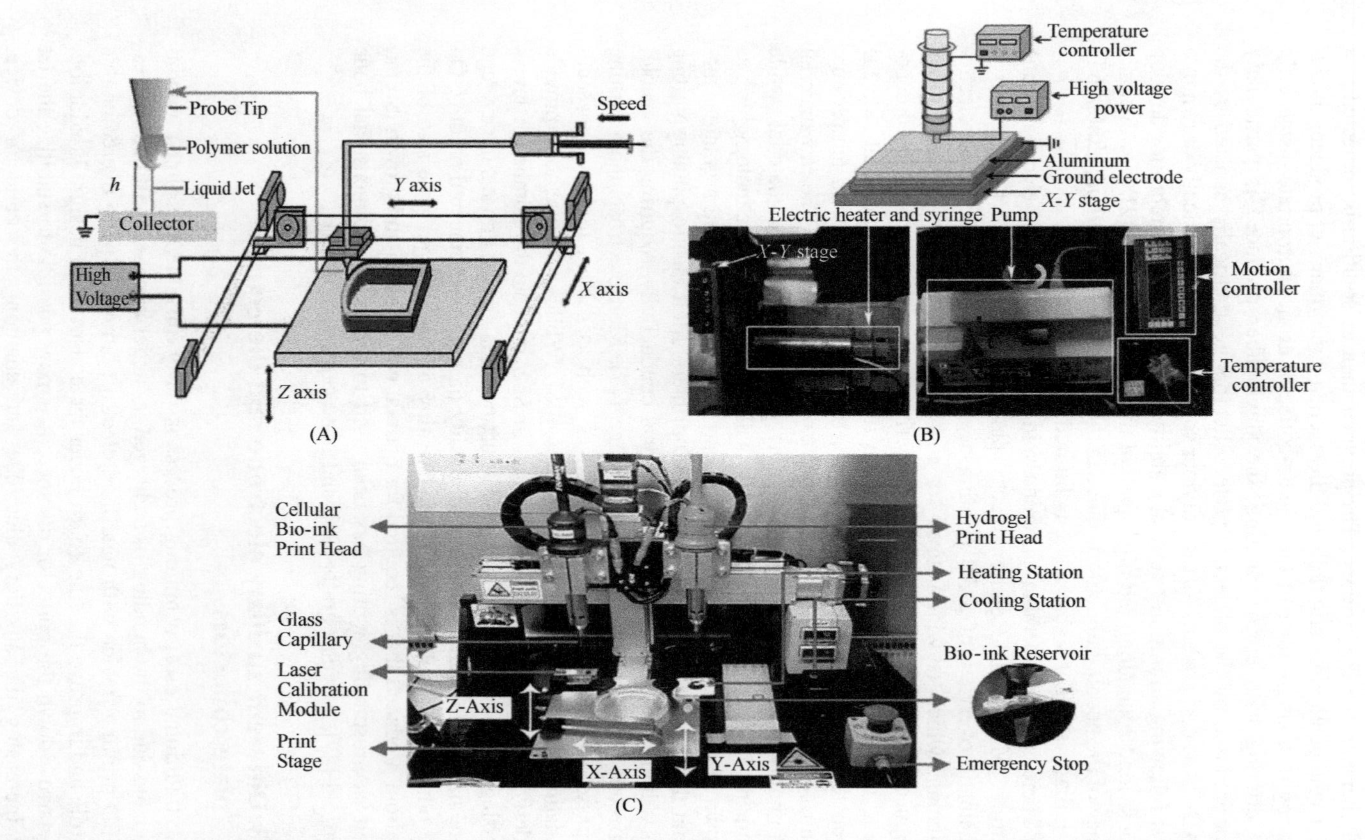

FIGURE 8.1 Schematic representation of a 3D printing device with (A) solution electrospinning [13], (B) melt electrospinning, and (C) fused deposition modeling [43].

This, in turn, affects the electric field between the nozzle and the collector plate that optimally guides the fiber in layers, in a controlled fashion. The resultant scaffold has enough internal space and a larger pore size that resembles the extracellular microenvironment of tissue despite a complex exterior contour [52].

8.3.1 Setup for electrospinning with an electrostatic lens system

Electrospinning with an electrostatic lens system has a characteristic annular auxiliary electrode in the form of a tube between the nozzle and the collector plate. Notably, the ring electrode, comprised of a single coil conductor, as shown in Fig. 8.2A and B, is subject to a voltage, which is of a lower value than the nozzle [53]. Neubert S et al., recently developed a distinct ring auxiliary electric circle to manufacture nylon, polyvinyl chloride (PVC), and polylactic acid (PLA). To this end, they achieved complex maneuvers with an average diameter of 0.15 mm and point positioning accuracy of 3 mm [53]. This highlights a novel method for preparing for tissue engineering scaffolds aimed at cellular differentiation. In contrast, Tansel et al., printed two different auxiliary electrodes [54]. While the cylindrical electrode was used in the first

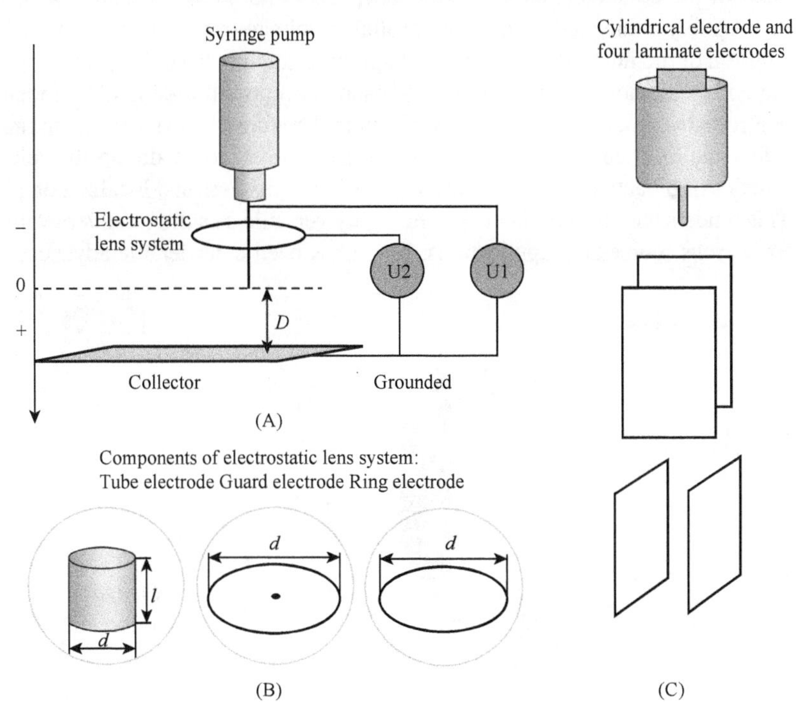

FIGURE 8.2 Schematic setup for electrospinning with (A) an electrostatic lens system, (B) components of the electrostatic lens system, and (C) a cylindrical electrode and four laminate electrodes [53,54].

study, the second design emphasizes the use of both cylindrical electrodes and four laminate electrodes, as highlighted in Fig. 8.2C. Their study shows different patterns on polyvinyl acetate (PVA) nanofiber mats that significantly enhance the mechanical properties of the material. Such constructs are expected to find a wide range of applications, mainly in clinical-grade dental composites. Researchers can consider the effects of unusual-diameter ring coils and different voltages while making tissue engineering scaffolds.

The auxiliary electrode offers many advantages over traditional electrospinning. Firstly, it enhances the precision of point positioning. Secondly, a staged electric field is used to increase the collection distance. However, this device has a stringent requirements for construction.

8.3.1.1 A core–shell nozzle

The core–shell or coaxial nozzle is a concentric set of coaxial spinnerets composed of an outer and inner needle, as illustrated in Fig. 8.3A and B. Coaxial electrospinning can generate fibers from various solution pairs, such as, core-sheath, hollow, and functional fibers that may contain particles or bioactive agents to accelerate the growth and differentiation of cells [53,55]. Both of the composite structures of the nozzle, the inner and outer shell, are respectively connected with the controllable voltage of the same polarity. The solution of the internal and external channels, although dissimilar, converges at the end of the nozzle. The flow rate of each concentric nozzle is independently controlled to produce construction with optimal features. Coaxial electrospinning offers multiple compartments within the same fiber, to modulate the release kinetics of bioactive agents, by altering the fiber thickness and localization [57]. This renders the resultant fiber system highly versatile as a drug-delivery vehicle for various biomedical applications. There has been considerable advancement

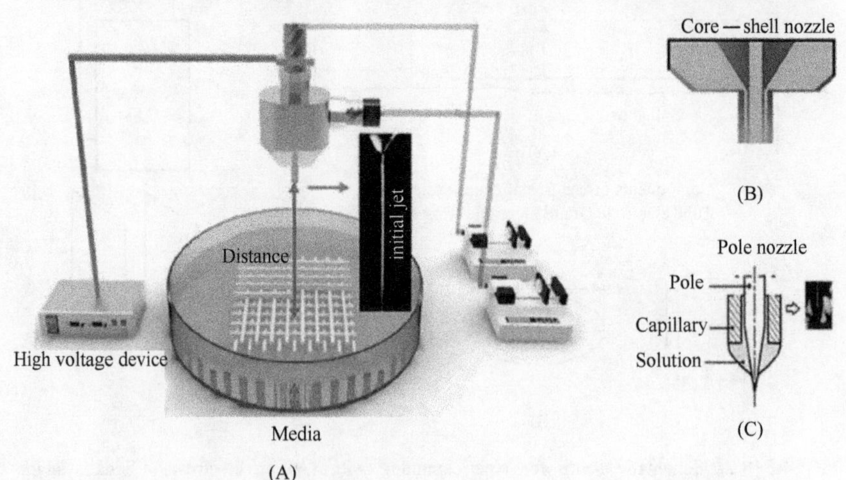

FIGURE 8.3 Schematic device of (A) near-field electrostatic spinning [55], (B) core–shell nozzle [55], and (C) a pole nozzle [56].

in the modeling of drug release from multiaxial electrospun fibers in drug delivery for tissue regeneration applications. Recently, Kim and team [55] proved that the printed coaxial microporous superfine fiber of the polycaprolactone (PCL)/collagen scaffold had better mechanical properties than the fibers prepared by a standard electrospinning process (greater than three times). An additional coating of type I collagen further improved the ability of proliferation and differentiation of osteoblast cells (MC3T3-E1 cells) in vitro. Overall, different physical and mechanical properties of the absorption of water and protein shows better cell activity than traditionally electrospun fibers.

Coaxial electrospinning resolves the problem of limitations in traditional drug-delivery methods. Unlike traditional methods, the coaxial method protects the therapeutic agent from the surrounding environment. It further helps in controlled drug release that maintains the blood level of a drug between the toxic concentration and minimum threshold concentration for an extended period. Three-dimensional printing of coaxial spinning for regeneration is based on the concept of designing optimal fiber design, such as a hollow or drug-filled core to develop functional substitutes that restore, improve, or repair damaged tissues and/or organs [58–60]. The simplistic setup can be employed on an industrial scale without compromising the structural, functional, and mechanical integrity of the material.

8.3.1.2 A pole nozzle

The key feature of a pole nozzle, illustrated in Fig. 8.3C, is the solid core that serves as a conductive tip [56]. The remainder of the device resembles a traditional 3D printer with electrospinning. The needle core is robust and allows flexibility in elongation. As the auxiliary electric field is formed at the needle core it decreases the average voltage of the jet. This forms the underlying principle of the device since the decrease in voltage plays a critical role in jet stabilization.

Lin and group have demonstrated the application of a core nozzle to pattern direct-writing of polyethylene oxide (PEO) [61]. Firstly, their research shows that the swing amplitude was considerably smaller than noncore nozzle direct-writing used in a traditional method. Secondly, the voltage required as well as the jet oscillation amplitude are significantly lower. In a parallel study, they shows an enhancement in the overall stability of a single fiber produced through a core nozzle and the induced type device. Owing to the stability of the jet, the fibers obtained are uniform and accurate in size. The fibers were of micro/nanoscale magnitude with a spiral structure fiber that may be applied as a biological scaffold for drug-delivery applications [13,61].

8.3.1.3 Other auxiliary electric fields and nozzles

Apart from the auxiliary electric field and nozzle described above, parallel plate auxiliary electric fields [62] and hole auxiliary electric fields [63] are

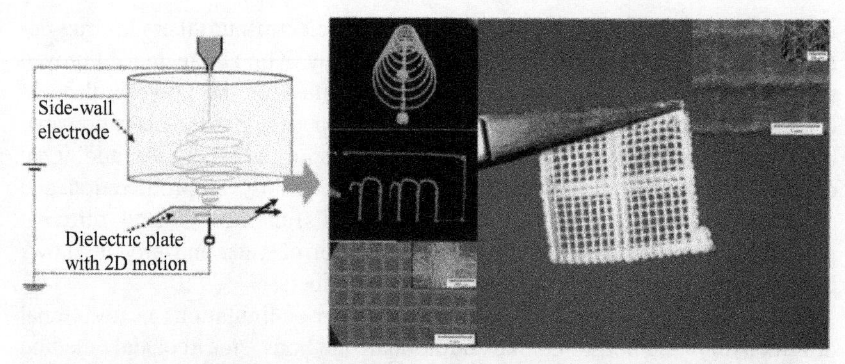

FIGURE 8.4 Three-dimensional printer with sharp-pin electrode dielectric plate [64].

commonly used. The parallel plate auxiliary electric field has the capacity to increase the flight distance of the jet, making it a suitable choice for large-scale production. On the other hand, the whole auxiliary electric field has the highest neutral requirement for the jet. These auxiliary electric fields and nozzles are rarely used for biological scaffold production.

8.3.2 Dielectric plate with sharp-pin electrode

The dielectric plate, illustrated in Fig. 8.4, has a sharp pin electrode that induces the formation of a concentrated electric field in the lower part [64]. This method is often combined with other auxiliary electric fields, such as the annular auxiliary electric field, to exert control over the spinning fiber. Lee and team [64] proved the feasibility of this combination device to produce a scaffold aimed particularly at cell patterning. In recent years, this method has gained considerable popularity and has also served as the motivation for the invention of the array dielectric collector.

8.4 Patterned, tubular, and porous nanofiber

There has been significant focus on biopolymers with micro/nanofibers in the field of tissue engineering and regenerative medicine [65]. Significant interdisciplinary research has shown their tremendous potential as wound dressings [66], biological aerosol filters [67], drug carriers [68], as well as biological imaging aids for artificial organs. This is attributed to the tunability of the fibers according to the desired tissue of the application. In order to successfully create a biological substituent, it is desirable that the scaffold mimics the biomechanics of the tissue microenvironment. Patterned controlled deposition of fibers is one such commonly used method for 3D scaffolds that enables biomechanical tunability by optimizing fiber layer control. It addresses the impediment imposed by the dense arrangement of fibers in the traditional electrospinning process. Yuan and team [69] directly printed and patterned polytechnic acid three-dimensional ultrasound fibrous scaffolds. Stable jet printing allows ultrafine prostatic acid fibers, which facilitated vascular smooth muscle cells, to penetrate and grow within the internal space of the scaffold. Such 3D growth of cells within scaffolds has been a major hurdle in the medical translation of electrospun scaffolds. Such an advancement is now leading to a 3D printer for electrospinning gradually replacing traditional electrospinning.

Fig. 8.5 depicts the preparation of tubular fibers by a 3D printer with electrospinning. Although traditional methods of patterning are associated with a lack of orderly fibers, the direct writing method eliminates most such organizational problems. Jana and team demonstrated the production of these tubular constructs with one natural and two synthetic polymers [chitosan, polycaprolactone, and poly(vinyl pyrrolidone)] [70]. They further demonstrated in a model bioapplication that muscle cells, cultured on the inner surface of an aligned nanofibrous tubular scaffold, enabled the formation of aligned and densely populated myotubes, organized as in natural muscle tissues.

Porous fibers prepared by a 3D printer based on electrospinning have advantages over traditional fibers, such as: (1) orderly and controlled fibers with large fiber diameter; (2) superior biocompatibility; and (3) superior cell permeability within the scaffold. This imparts biomimetic character to the scaffold, which resembles the extracellular matrix comprising of cells within the structural compartments. In nature, the extracellular matrix is composed of 3D structural components in micro/nanoscale that form a supportive meshwork around cells [15]. In order to produce successful tissue engineered constructs, it is desirable to have a 90% porous environment, which can mitigate cell adhesion as well as the infiltration of nutrients and metabolic products [71]. The control over the microstructure offered by 3D printing recapitulates the natural form and function of the tissue microenvironment through its high surface area to volume ratio and porosity.

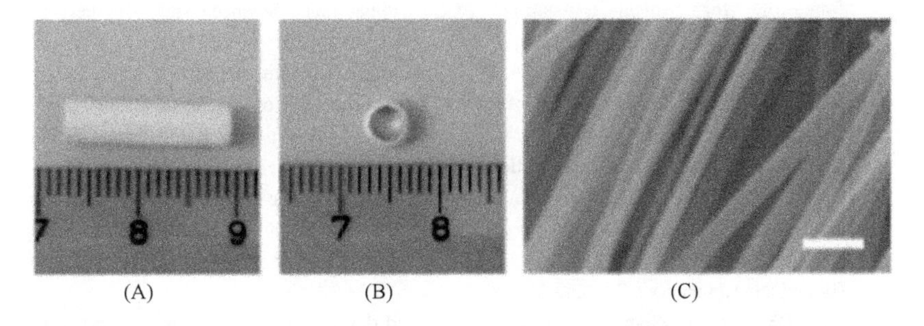

FIGURE 8.5 Tubular nanofibers prepared by electrospinning polyvinylpyrrolidone (PVP) solution. (A) Photograph of an aligned fibrous tubular scaffold. (B) A cross-sectional view of the tube. (C) SEM image of the fibrous structure of aligned tubular constructs showing fiber diameters of 150–250 nm [70].

8.5 Conclusion

The 3D printer with electrospinning technology is the amalgamation of traditional near-field electrospinning technology and advanced 3D printing technology. Recent advances in this field have helped to overcome the limitations of traditional electrostatic spinning, mainly in terms of control of fiber deposition. Novel methods in scaffold fabrication methods are based on complex external geometries, yet they offer improved permeability of cells in the internal space, increasing the overall quality of the surgical construct. Although in its infancy, pioneering research into 3D printing with electrospinning shows great promise in the field of tissue engineering and regenerative medicine. With such technology in place, it is now possible to deliver a personalized surgical construct for treating tissue effects that mimic the tissue microenvironment and release bioactive agents and drugs to the desired site.

Currently, there are limited equipment options on the market that allow researchers to assemble their setup manually. However, the field continues to evolve with different process mechanics being optimized to target specific areas of biomedical research. Overall, a significant and promising path lies ahead for multidisciplinary research, combining materials engineering, computational modeling, medical imaging, and cell biology, to fully achieve the potential of 3D printing electrospinning in regenerative medicine.

References

[1] Pina S, Oliveira JM, Reis L. Natural-based nanocomposites for bone tissue engineering and regenerative medicine: a review. Advanced Materials 2015;27(7):1143–69.

[2] Huyer LD, Zhang B, Korolj A, et al. Highly elastic and moldable polyester biomaterial for cardiac tissue engineering applications. ACS Biomaterials Science and Engineering 2016;2:780–8.

[3] Hollister SJ, Flanagan CL, Morrison RJ, Patel JJ, Wheeler MB, Edwards SP. Image-based design and 3d biomaterial printing to create patient specific devices within a design control framework for clinical translation. ACS Biomaterials Science and Engineering 2016;2(10):1827–36.

[4] Shelemin A, Nikitin D, Choukourov A, et al. Preparation of biomimetic nano-structured films with multi-scale roughness. Journal of Physics D: Applied Physics 2016;49(25):254001.

[5] Low KH, Hu T, Mohammed S, et al. Perspectives on biologically inspired hybrid and multi-modal locomotion. Bioinspiration and Biomimetics 2015;10(2):020301.

[6] Shao W, He J, Han Q, et al. A biomimetic multilayer nanofiber fabric fabricated by electrospinning and textile technology from polylactic acid and Tussah silk fibroin as a scaffold for bone tissue engineering. Materials Science and Engineering: C Materials for Biological Applications 2016;67:599–610.

[7] Song QS, Liu YX, Zhang YC, et al. Research and development in electrospinning theory and technology. Materials Science Forum 2015;815:695–700.

[8] Bhardwaj N, Kundu SC. Electrospinning: a fascinating fiber fabrication technique. Biotechnology Advances 2010;28(3):325–47.

[9] Holzwarth JM, Ma PX. Biomimetic nanofibrous scaffolds for bone tissue engineering. Biomaterials 2011;32(36):9622–9.

[10] Jiang T, Carbone EJ, Lo WH, et al. Electrospinning of polymer nanofibers for tissue regeneration. Progress in Polymer Science 2015;46:1–24.

[11] Fuh YK, Wang BS, Liu BJ. Near-field sequentially electrospun three-dimensional piezoelectric fibers arrays for self-powered sensors of human gesture recognition. Nanomaterials and Energy 2016;30:677–83.

[12] Hao P, Zhang J, Li X, et al. Modes of centrifugal electrospinning. Engineering Plastics Application; 2015.

[13] Sun D, Chang C, Li S, et al. Near-field electrospinning. Nano Letters 2006;6(4):839.

[14] Brown TD, Dalton PD, Hutmacher DW. Direct-writing by way of melt electrospinning. Advanced Materials 2011;23(47):5651–7.

[15] Yu YZ, Zheng LL, Chen HP, et al. Fabrication of hierarchical polycaprolactone/gel scaffolds via combined 3D bioprinting and electrospinning for tissue engineering. Advances in Manufacturing 2014;2(3):231–8.

[16] Seyednejad H, Gawlitta D, Kuiper RV, et al. In vivo biocompatibility and biodegradation of 3D-printed porous scaffolds based on a hydroxyl-functionalized poly(ε-caprolactone). Biomaterials 2012;33(17):4309–18.

[17] Nair K, Gandhi M, Khalil S, et al. Characterization of cell viability during bioprinting processes. Biotechnology Journal 2010;4(8):1168–77.

[18] Dombrowski F, Caso PWG, Laschke MW, et al. 3-D printed bioactive bone replacement scaffolds of alkaline substituted ortho-phosphates containing meta- and di-phosphates. Key Engineering Materials 2012;529–530(1):138–42.

[19] Ozbolat IT, Yu Y. Bioprinting toward organ fabrication: challenges and future trends. IEEE Transactions on Biomedical Engineering 2013;60(3):691–9.

[20] Murphy SV, Atala A. 3D bioprinting of tissues and organs. Nature Biotechnology 2014;32(8):773–85.

[21] Ravichandran R, Venugopal JR, Mueller M. Buckled structures and 5-azacytidine enhance cardiogenic differentiation of adipose-derived stem cells. Nanomedicine 2013;8(12):1985–97.

[22] Zhang T, Yao R, Sun W, et al. Research on the path of biological 3D printing industry in Beijing City. New Mater Indus 2013;8:25–30.

[23] Chen X. The development and application of 3D printing technology in medical. Guangdong Sci Tech 2014;15:60–3.

[24] Wang XM. New type industrialization, biomaterials. J New Indus 2015;12:37–68.

[25] Musong S, Junxia G, Wenzhi S, et al. Application of 3D printing technology in biomedical field. World Journal of Complex Medicine 2015;5:57–8.

[26] Zhang HB, Hu DC. New research and applications of bio-3D printing. Powder metal indus 2015;4:63–7.

[27] Zhang W, Lian Q, Li D, et al. Cartilage repair and subchondral bone migration using 3D printing osteochondral composites: a one-year-period study in rabbit trochlea. BioMed Research International 2014;2014(5):746138.

[28] Rondinoni C, Grillo FW, Matias CM, et al. Use of 3D printing in surgical planning: strategies for risk analysis and user involvement. Proceedings - IEEE Symposium on Computer-Based Medical Systems 2015:29–30.

[29] Scheidler J, Hricak H, Vigneron DB, et al. Prostate cancer: localization with three-dimensional proton MR spectroscopic imaging—clinicopathologic study. Radiology 1999;213:473–80.

[30] Mironov V, Boland T, Trusk T, et al. Organ printing: computer-aided jet-based 3D tissue engineering. Trends in Biotechnology 2003;21(4):157−61.

[31] Yu-Hui H, Rosemary S, Linping Z, et al. Virtual surgical planning and 3D printing in prosthetic orbital reconstruction with percutaneous implants: a technical case report. International Medical Case Reports Journal 2016;9:341−5.

[32] Yu Y, Hua S, Yang M, et al. Fabrication and characterization of electrospinning/3d printing bone tissue engineering scaffold. RSC Advances 2016;6:110557−65.

[33] Ma YM, Liu YY, Chen HP, et al. Preparation and characterization of tissue engineering scaffolds by composite molding of 3D printing and electrospinning. Key Engineering Materials 2015;645−646:1368−73.

[34] Agarwal S, Wendorff JH, Greiner A. Use of electrospinning technique for biomedical applications. Polymer 2008;49(26):5603−21.

[35] Min BM, Lee G, Kim SH, et al. Electrospinning of silk fibroin nanofibers and its effect on the adhesion and spreading of normal human keratinocytes and fibroblasts in vitro. Biomaterials 2004;25(7):1289−97.

[36] Cato L, Sangamesh K, Syam N, et al. Recent patents on electrospun biomedical nanostructures: an overview. Recent Patents on Biomedical Engineering 2008;1(1):68−78.

[37] Sill TJ, Recum HAV. Electrospinning: applications in drug delivery and tissue engineering. Biomaterials 2008;29(13):1989−2006.

[38] Goldstein AS, Thayer PS. Chapter 11 − fabrication of complex biomaterial scaffolds for soft tissue engineering by electrospinning. Nanobiomaterials in Soft Tissue Engineering 2016:299−330.

[39] Khil MS, Cha DI, Kim HY, et al. Electrospun nanofibrous polyurethane membrane as wound dressing. Journal of Biomedical Materials Research Part B: Applied Biomaterials 2010;67(2):675.

[40] Wang ZG, Wan LS, Liu ZM, et al. Enzyme immobilization on electrospun polymer nanofibers: an overview. Journal of Molecular Catalysis B: Enzymatic 2009;56(4):189−95.

[41] Tran DN, Balkus Jr KJ. Enzyme immobilization via electrospinning. Topics in Catalysis 2012;55(16−18):1057−69.

[42] Shin YM, Hohman MM, Brenner MP, et al. Electrospinning: a whipping fluid jet generates submicron polymer fibers. Applied Physics Letters 2001;78(8):1149−51.

[43] Yu D, Li X, Wang D. Process analysis and application for rapid prototyping based on fused deposition modeling. Mac Des Manuf 2011;8:65−7.

[44] Qiang Y, Jing T. Application of three-dimensional printing technique in manufacturing scaffolds for bone tissue engineering. Chinese Journal of Tissue Engineering Research 2015;30:4870−5.

[45] Pelcl M, Chvojka J, Kuželová E,K. The combination of 3D printing and nanofibers for tissue engineering of articular cartilage. Nano 2014;11:1−6.

[46] Jiang H, Yikang W, Xiaoqing T, et al. Design and study of fused deposition modeling(FDM) 3D printing process parameters optimization. Manufacturing Technology and Machine Tool 2016;6:139−42.

[47] Lan H, Li D, Lu B. Micro-and nanoscale 3D printing. Chinese Space Science Technology 2015;9:919−40.

[48] Brown TD, Edin F, Detta N, et al. Melt electrospinning of poly(ε-caprolactone) scaffolds: phenomenological observations associated with collection and direct writing. Materials Science and Engineering C 2014;45:698−708.

[49] Martins A, Chung S, Pedro AJ, et al. Hierarchical starch-based fibrous scaffold for bone tissue engineering applications. Journal of Tissue Engineering and Regenerative Medicine 2010;3(1):37–42.
[50] Centola M, Rainer A, Spadaccio C. Combining electrospinning and fused deposition modeling for the fabrication of a hybrid vascular graft. Biofabrication 2010;02:1067–75.
[51] Yeo M, Lee H, Kim G. Three-dimensional hierarchical composite scaffolds consisting of polycaprolactone, β- tricalcium phosphate, and collagen nanofibers: fabrication, physical properties, and in vitro cell activity for bone tissue regeneration. Biomacromolecules 2011;12:502–10.
[52] Liu Y, Zhang F, Chen W, et al. CAD/CAM system and experimental study of biological 3D printing composite process. Journal of Mechanical Engineering 2014;50:147–54.
[53] Neubert S, Pliszka D, Gora A, et al. Focused deposition of electrospun polymer fibers. Journal of Applied Polymer Science 2012;125:820–7.
[54] Uyar T, Çökeliler D, Doğan M, et al. Electrospun nanofiber reinforcement of dental composites with electromagnetic alignment approach. Materials Science and Engineering C 2016;62:762–70.
[55] Kim MS, Kim GH. Electrohydrodynamic direct printing of PCL/collagen fibrous scaffolds with a core/shell structure for tissue engineering applications. Chemical Engineering Journal 2015;279:317–26.
[56] Xu L, Wang C. Impact of blow heads with electric fluid coupled spray printing. J Wuhan Eng Inst 2013;25:11–5.
[57] Khalf A, Madihally SV. Recent advances in multiaxial electrospinning for drug delivery. European Journal of Pharmaceutics and Biopharmaceutics 2016;112:1–17.
[58] Sun Z, Zussman E, Yarin AL, et al. Compound core–shell polymer nanofibers by Co-electrospinning. Advanced Materials 2010;15(22):1929–32.
[59] Park JH, Braun PV. Coaxial electrospinning of self-healing coatings. Advanced Materials 2010;22(4):496–9.
[60] Hong-Jun L, Yong H. Fabrication and mechanical properties of PAN/PEO and CA/PEO coaxial electrospun fibers. Polymer Bulletin 2013;41(6):64–70.
[61] Liu K, Long Y, Tang C, Yin H, Cao K, Yin Z, Li M. Recent advance in fabrication of helical structural nanofibers via electrospinning. Materials Review 2010;24:72–6.
[62] Zhou FL, Gong RH, Porat I. Three-jet electrospinning using a flat spinneret. Journal of Materials Science 2009;44(20):5501–8.
[63] Boland T, Tao X, Damon BJ, et al. Drop-on-demand printing of cells and materials for designer tissue constructs. Materials Science and Engineering C 2007;27(3):372–6.
[64] Lee J, Lee SY, Jang J, et al. Fabrication of patterned nanofibrous mats using direct-write electrospinning. Langmuir 2012;28(18):7267–75.
[65] Yang Y, Li L. Application of natural artificial synthetic polymer hybrid nanofibers in biomedical applications. Medical Biomechanics 2011;26:105–8.
[66] Unnithan AR, Barakat NAM, Pichiah PBT, et al. Wound-dressing materials with antibacterial activity from electrospun polyurethane–dextran nanofiber mats containing ciprofloxacin HCl. Carbohydrate Polymers 2012;90(4):1786–93.
[67] Wang R, Wang Z, Wu J, Wang T. The application of electrospinning nanofibers and their application in biological aerosol filtration. Synth Fiber Ind 2013;4:52–5.
[68] Wang Z. Construction of drug-loaded electrospun organic/inorganic hybrid nanofibers biomedical applications. Materials China 2014;11:661–8.

[69] Yuan H, Zhou Q, Li B, et al. Direct printing of patterned three-dimensional ultrafine fibrous scaffolds by stable jet electrospinning for cellular ingrowth. Biofabrication 2015;7(4):045004.
[70] Jana S, Zhang M. Fabrication of 3D aligned nanofibrous tubes by direct electrospinning. Journal of Materials Chemistry B 2013;1(20):2575−81.
[71] Liu HL, Gu HQ. Natural extracellular matrix and its application in tissue engineering. Cn J Dialysis Artif Organs 2006;17:42−6.

Chapter 9

Fiber membranes obtained by melt electrospinning for drug delivery

Chapter outline

9.1 Introduction

Controlled drug-delivery systems offer numerous advantages over conventional dosage forms, including improved therapeutic effects, reduced toxicity, and increased patient compliance and convenience [1]. In such systems, synthetic, natural, and hybrid polymer materials are used as drug vehicles to enhance the efficiency of the system [2,3]. Polyhydroxybutyrate (PHB) is a biocompatible, nontoxic, and biodegradable polymer obtained from a natural source that is suitable for biomedical applications [4]. This polymer is one of the most promising biomedical materials because of its appropriate biocompatibility and controlled biodegradation. PHB has a high degree of crystallinity because of its perfect stereoregularity and high purity. However, the excessive brittleness, poor processability, and poor thermal stability of PHB limit its potential applications as a pristine material [5]. Multicomponent fibers have attracted increasing attention because new properties can be obtained by combining different materials. We combined PHB with other biopolymers to

Melt Electrospinning. https://doi.org/10.1016/B978-0-12-816220-0.00009-9

create a drug-delivery system to investigate the diffusion transport kinetics of drugs. Poly-L-(lactic acid) (PLLA) was chosen, because it has been widely used in various biomedical applications [6–8], based on its biodegradability, biocompatibility, and good mechanical properties.

The unresolved problems for polymeric drug-delivery systems include their low efficiency for preparing nano- and microparticles or vesicles, and their low efficiency for drug delivery [9]. Electrospinning is a simple, low-cost, and versatile process for preparing a wide range of polymeric microfibers and nanofibers in a controllable manner [10]. Moreover, this technique is very convenient because the fiber can be coated onto any surface with nonwoven or aligned fibers. Electrospinning can be generally divided into two types of methods: solution electrospinning and melt electrospinning. Solution electrospinning is usually used to prepare nanofibers because it uses simple equipment and produces thinner fibers than melt electrospinning. However, many of the residual solvents in the fibers are toxic to cells and tissues [11]. In contrast, melt electrospinning provides a solvent-free process to fabricate polymer fibers using a safe, eco-friendly method. Thus, the disadvantages of solution electrospinning, such as expensive solvent recovery, the risk of solvent explosion, and residual solvent toxicity are not applicable. These features are becoming increasingly important in the pharmaceutical industry.

The chosen drug, dipyridamole (DPD), is an antithrombotic and antithrombogenic drug. It can reduce smooth muscle cell proliferation and promote vascular endothelial cell proliferation. In addition, DPD has been employed with acetylsalicylic acid for secondary stroke prevention because of its antithrombotic effect [12]. DPD had been blended with different polymers and then electrospun into a biodegradable fibrous scaffold or other forms for drug delivery [13–15]. However, the toxic solvent residue in the fibers cannot be eliminated because it must be used to dissolve the polymer. Thus, in this study solvent-free melt electrospinning was applied to prepare novel drug-delivery systems.

Controlled drug release from nondegradable polymer systems occurs through a diffusion mechanism [16–18], while degradable polyesters, such as PLLA and PHB, with unstable chemical bonds, can be affected by hydrolytic reactions, specifically through the mechanism of end-group autocatalysis [19]. In this connection, despite the impressive technological achievements in the creation of ultrathin fibers, an extremely limited amount of work [20–22] has been devoted to diffusion and hydrolytic problems that emerge in fibrillar materials during drug delivery. These problems require comprehensive experimental and theoretical examinations. To solve the coherent diffusion-kinetic problems, measurements of controlled drug release from macroscopic fibrillar membranes (the mats) under various intrinsic conditions must be executed.

In the present work, fiber membranes with various proportions of PLLA/PHB were produced by melt electrospinning. The developed materials were morphologically, structurally, and thermally characterized. PLLA/PHB fiber membranes without any toxic solvents and containing different amounts of DPD were prepared to assess the polymeric drug-delivery system. The influence of DPD on the fiber diameter, and the structural and thermal properties, was characterized. Further, drug encapsulation in the fibers was confirmed. The dynamic behavior of the polymer molecules was evaluated by their segmental mobility and drug diffusion, which is the most important process for controlled drug release.

9.2 Experimental

9.2.1 Materials

PHB (1001MD, BASF) and PLLA particles ($M_n = 100{,}000$; Zhejiang Hisun Biomaterials Co. Ltd., China) were dried at 60°C for 6 h in a vacuum oven before the experiments. DPD ($M_w = 504.63$; >98.0%) was purchased from Beijing Inoke Technology Co. Ltd., China.

9.2.2 Processing of the blends

Blends of PLLA and PHB at different PLLA/PHB ratios of 9:1, 8:2, 7:3, and 6:4 by weight were mixed in a Haake poly lab torque rheometer at 190°C for 6 min at a constant rotor speed of $80\,r\cdot min^{-1}$. The PLLA/PHB ratios of 9:1 and 7:3 with 1% and 5% DPD by weight have the same conditions except for the temperature at 170°C.

9.2.3 Melt electrospinning

The melt electrospinning devices used were the same as those in our previous studies [23,24]. The collector covered with aluminum foil was connected to the positive of the power, and the spinneret was grounded. Samples were prepared using a normal power supply [DW-P503-2ACDE; Dong Wen high-voltage power supply (Tianjin) Co. Ltd., China]. First, several electrospinning processing conditions were tried to produce the electrospinning fiber of PLLA at the ratio of 9:1. The different positive voltages, applied cylinder temperature, and spinning distance were investigated for the formation of a stable Taylor cone to obtained a uniform and continuous fiber. Because of the addition of the DPD, the applied temperature could be decreased notably, and it was fixed to 170°C. The DPD acts as the plasticizer in melt electrospinning and decreases polymer–polymer interaction to make segments of the polymer move at low temperature [25]. Thus the melt electrospinning was carried out at 220°C without DPD and 170°C with DPD, the applied voltage was 35 kV, and the distance between the collector and spinneret was set at 7 cm. A schematic diagram of the experimental devices and the spun fiber of 7:3 with 1% DPD is shown in Fig. 9.1.

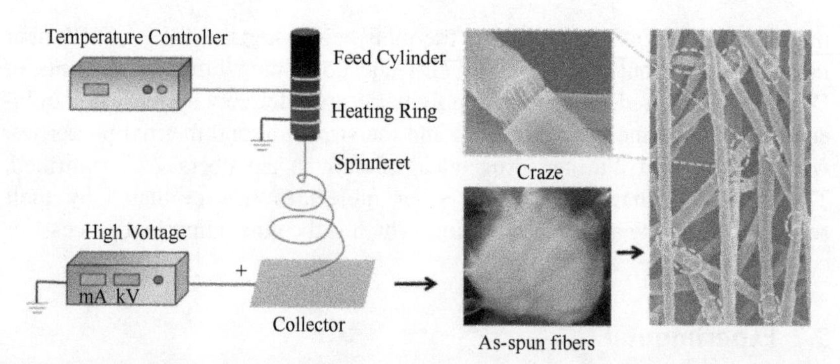

FIGURE 9.1 Schematic diagram of the experimental devices and the as-spun PLLA/PHB fibers of 7:3 with 1% DPD [26].

Electrospun fiber membranes were coated with gold and their morphology was characterized using scanning electron microscopy (SEM; Hitachi S4700, Japan) with an accelerating voltage of 20 kV. The average fiber diameter and its size distribution were calculated using SEM images by Image J software. Over 20 fibers of each sample were randomly selected from the SEM image, and each fiber was measured at five different locations. The crystalline phases of the electrospun fibers were performed using X-ray diffraction (XRD) (D8, Bruker, Germany), in the 2θ range 5°−90° with Cu Kα radiation ($\lambda = 1.5406$ Å), with a voltage and current of 40 kV and 40 mA, respectively. The equipment resolution and the scan speed were 0.02° and 0.1 $s \cdot step^{-1}$, respectively. Thermophysical characteristics and crystalli zation behavior were obtained by differential scanning calorimetry (DSC; PerkinElmer Pyris 1) with the samples being heated from 0 to 200°C at 10°C·min^{-1} under a nitrogen atmosphere (20 mL · min^{-1}). The glass transition temperature (T_g), cold crystallization temperature (T_c), melting temperature (T_m), cold crystallization enthalpy (ΔH_c), and melting enthalpy (ΔH_m) were obtained from the first heating, and the degree of crystallinity X_c (%) was calculated using Eq. (9.1) [27]:

$$X_c = \left[\frac{\Delta H_m - \Delta H_c}{\Delta H_f}\right] \times \frac{1}{W_{PLLA}} \times 100\% \tag{9.1}$$

ΔH_f is the melting heat associated with pure crystalline PLLA $(93\text{J}\cdot\text{g}^{-1})$ [28] and $1/W_{PLLA}$ is the proportion of PLLA in the blend. Infrared measurements [Fourier transformed infrared spectroscopy (FTIR)] were performed at room temperature with a Thermo Electron 8700 apparatus in transmission mode from 4000 to 500 cm^{-1}. FTIR spectra were collected after 10 scans with a resolution of 4 cm^{-1}. The measurements were performed with potassium bromide pellets (KBr). Segmental mobility of PLLA and PHB in the blend fibers was studied by the probe electron spin-resonance spectroscopy (ESR) method. X-band EPR spectroscopy X-band electron paramagnetic resonance spectroscopy (EPR) were registered on an automated EPR-V spectrometer (Semenov Institute of Chemical Physics, Moscow RF.). A stable nitroxide radical TEMPO (2,2,6,6-tetramethylpyperidin-1-oxil) was used as a probe. The radical was introduced into the fibers from the gas phase at 40°C. The radical concentration in the polymer was not above 10^{-3} $mol \cdot L^{-1}$. The experimental spectra of the spin probe in the region of slow motions ($\tau > 10^{-10}$s) were analyzed within the model of isotropic Brownian rotation using a program described by Budil et al. [29]. The spectra were modeled using the following main values of the g-tensor and the hyperfine coupling tensor of the radical: $g_{xx} = 2.0096$, $g_{yy} = 2.0066$, $g_{zz} = 2.0025$, $A_{xx} = 7.0$ G (1G=10^{-4} T), $A_{yy} = 5.0$ G, and $A_{zz} = 35.0$ G. The value A_{zz} is determined experimentally from the EPR spectra of the nitroxide radial in the polymer at −216°C; it is almost equal to the values that Timofeev et al. reported [30]. The correlation time of probe rotation, τ, is determined from the ESR spectra via the equation [31]:

$$\tau = \Delta H_{+} \times \left[(I_{+}/I_{-})^{0.5} - 1 \right] 6.65 \times 10^{-10} [\mathrm{s}] \tag{9.2}$$

where, ΔH_{+} is the width of spectrum component located in a weak field and I_{+}/I_{-} is the ratio of component intensity in weak and strong fields of the spectrum, respectively. The statistical error of τ measurements is equal to ±5. The kinetics of DPD release is studied with a Beckman DU-65 UV spectrophotometer (USA). In kinetic measurements, the fiber sample was immersed in an aqueous medium and the optical density of DPD samples was determined with periodic sampling. The interval of sampling depended on the fiber composition and, accordingly, on the rate of drug release; it is from 1 to 30 min. The experiments lasted from several tens of minutes to several hours. In the DPD ultraviolet (UV) spectra, there are two characteristic peaks at $\lambda = 410$ nm and a more intense peak at $\lambda = 292$ nm with an extinction coefficient of 31,260 $L \cdot mol^{-1} \cdot cm^{-1}$, which is used for the drug release measurement.

9.3 Results and discussion

9.3.1 Fiber membrane morphology

Scanning electron microscopy (SEM) micrographs of electrospun fibers in the membranes are shown in Fig. 9.2A–H. The fibers of the as-spun PLLA, PHB, and their blends (PLLA/PHB = 9:1 and 7:3) without DPD have smooth surfaces and relatively uniform fiber diameter distributions without any specific thickening, such as beads or spindle-like units. In contrast to the drug-unloaded fibers, the composite fibers with a PLLA/PHB ratio of 7:3% and 1% DPD have rough surfaces and nonuniform diameters within a single fiber. In this case, their morphology includes crazes that resemble lotus root crevices;

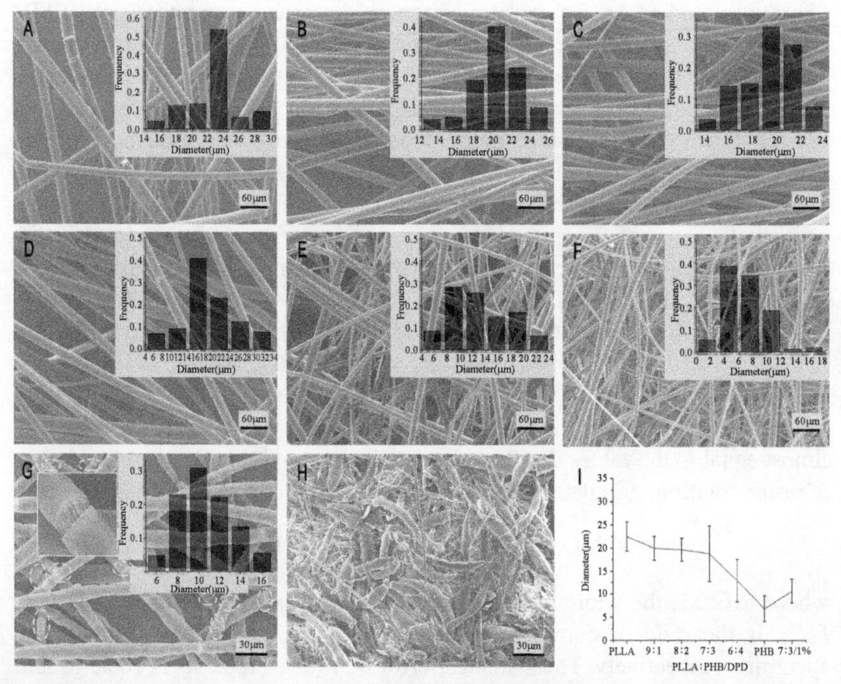

FIGURE 9.2 Morphology of the electrospun fiber membranes of: (A) pure PLLA and for PLLA/PHB ratios; (B) 9 : 1; (C) 8 : 2; (D) 7 : 3; (E) 6 : 4; (F) pure PHB; and 7 : 3 with (G) 1% and (H) 5% of DPD concentration; (I) the corresponding average fiber diameters [26].

the crazes are apparently disunited but remain connected with the nanofibrils. Analogous morphologies were discovered for the polymer samples under tensile testing [32,33]. However, we obtained similar structures without any intentional stretching during melt electrospinning.

In the present study, the crazes in the fibers are not evenly spaced along the fiber axis, and they are not present for some single filaments. In the absence of external mechanical stress and under identical electrospinning conditions for the nonloaded and drug-loaded samples, these fiber defects developed only in the presence of DPD and, hence, are directly related to the DPD molecule impact upon the fiber morphology and the segmental mobility of polymer molecules (see the ESR results). Generally, these effects are likely related to: (1) a slight polymer phase separation between the 7 : 3 polymer blends and DPD [34]; (2) electrically driven jet instabilities during electrospinning (nonaxisymmetric instability and two axisymmetric instabilities), which form a multipoint local stress concentration on the jet [33] and further expand the fluctuation; and (3) a decrease in the entanglement of molecular chains form with the addition of DPD, which contributes to the motion of the molecular chains. A further increase in the DPD concentration to 5% (w) causes

TABLE 9.1 PLLA/PHB membranes formulations and average fiber diameters [26].

Materials	PLLA (%(*w*))	PHB (%(*w*))	DPD (%(*w*))	Average fiber diameter (μm)
PLLA	100	–	–	22.53 ± 3.14
9:1	90	10	–	19.99 ± 2.56
8:2	80	20	–	19.69 ± 2.43
7:3	70	30	–	18.77 ± 6.02
6:4	60	40	–	12.96 ± 4.64
PHB	–	100	–	6.84 ± 2.77
7:3/1%	69.3	29.7	1	10.69 ± 2.51
7:3/5%	66.5	28.5	5	–

PLLA/PHB fiber disintegration. Fig. 9.2H shows that the samples are composed of short, crescent-like units that are very likely caused by jet discontinuity during electrospinning.

All electrospun fiber formulations and the corresponding average diameters are summarized in Table 9.1. The fiber diameter decreases as the PHB content increases in the composite fibers; for 7∶3 PLLA/PHB fibers, the fiber diameter reaches a minimum value of (18.8 ± 6.0) μm. The smallest diameter is observed for pristine PHB, as shown in Fig. 9.2I. DPD embedded in the polymer system decreases the mean fiber diameter to (10.7 ± 2.5) μm and simultaneously creates a narrower diameter distribution. The reduced fiber thickness from encapsulated DPD is previously described [35]; the authors proposed that ionic forms of DPD participate in the electrostatic interactions during fiber electrospinning. Simultaneously, the authors observed a surface tension reduction in the presence of DPD; this effect could cause fiber bending with the following discontinuity of the polymer jet [36]. The coherent consequence of this effect is observed in Fig. 9.2H for the 7∶3 compositions with 5% DPD loading.

9.3.2 Fourier transformed infrared spectroscopy

FTIR spectra of the PLLA, PHB, and PLLA/PHB membranes are shown in Fig. 9.3A. The band associated with the $C-CH_3$ stretching vibrations of PLLA appears at 1044 cm^{-1}, and its intensity decreases with the addition of PHB. The band at 1080 cm^{-1} is ascribed to the stretching vibrations of the C–O–C groups. In neat PHB, the band at 1271 cm^{-1} is ascribed to stretching vibrations

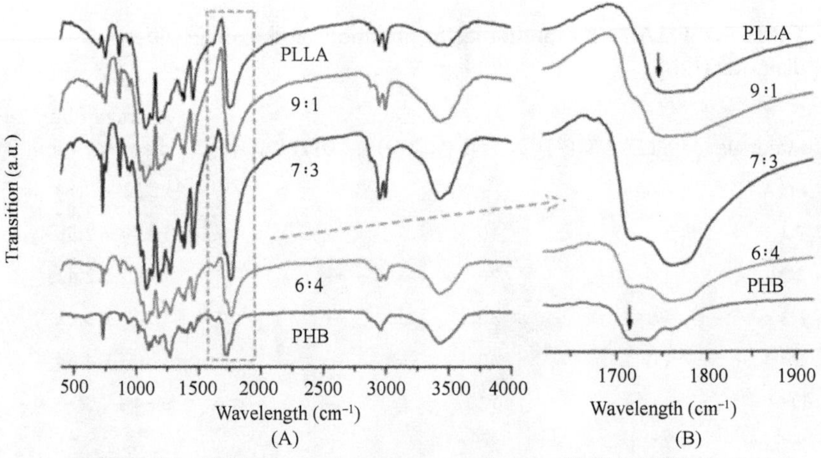

FIGURE 9.3 FTIR spectra of electrospun membranes with PLLA/PHB blends [26].

of the C–O–C groups, which decrease with the addition of PLLA [37]. At 1384 cm^{-1}, the CH_3 asymmetric deformation band appears, and its intensity increases with higher amounts of PHB [38]. In the spectrum of PLLA, the typical band at 1752 cm^{-1} is related to the crystalline carbonyl C=O vibration [39]. The amorphous carbonyl C=O vibration of PLLA is centered at 1745 cm^{-1}, but it is very weak and cannot be clearly identified [38]. In the spectrum of pure PHB, the bands at 1718 cm^{-1} and 1733 cm^{-1} are attributed to the stretching vibrations of crystalline carbonyl C=O groups, and a small shoulder at 1750 cm^{-1} contributes to the amorphous carbonyl C=O vibration [39,40]. For the PLLA/PHB blends, the carbonyl stretching bands of PLLA and PHB show different intensities with increasing PHB content. For the 6:4 blend, the band at 1722 cm^{-1} is ascribed to the stretching of the crystalline carbonyl group, and the band near 1733 cm^{-1} becomes narrower and overlaps with the amorphous carbonyl group of PLLA. In the 7:3 blend, two distinct carbonyl groups are observed; the band at 1752 cm^{-1} corresponds to the amorphous carbonyl vibration of PLLA, and the band at 1716 cm^{-1} corresponds to crystalline carbonyl stretching of PHB. However, for the 9:1 blend, the carbonyl band at 1745 cm^{-1} is broadened and shows a small shoulder near 1718 cm^{-1}. The different peak position in the blends is associated with a transesterification reaction between PLLA and PHB that forms intermolecular hydrogen [41].

The FTIR spectra show the typical characteristic vibrational bands of DPD at 1534 cm^{-1} corresponding to the C=N (ring) stretching vibration (Fig. 9.4) [42]. A new absorption band at 1582 cm^{-1} corresponding to the C=O stretching vibration appears, which may be associated with the formation of a carbonyl group during the melt electrospinning [43]. In this way, the presence of DPD is confirmed in the membranes after melt electrospinning.

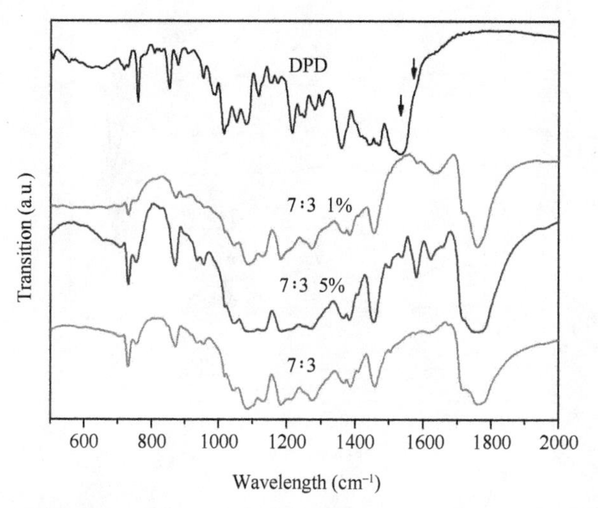

FIGURE 9.4 FTIR spectra of DPD and the ratio of 7:3 blend electrospun membranes with 0%, 1%, and 5% DPD [26].

9.3.3 Differential scanning calorimetry

The heating processes of PLLA, PHB, and their blends with and without DPD are observed in DSC scans (Fig. 9.5A). Pure PLLA has a glass transition temperature (T_g) of ~61°C and a sharp melting peak of ~173°C. For pure PHB, a small melting peak of 150°C is observed, and the glass transition is not discernible. Pure PHB shows a lower melting point than general because it has poor thermal stability in the melt and thermal degradation begins at the melting point [44].

For the PLLA/PHB blends, a glass transition is observed at ~57°C for the PLLA component, and the glass transition temperature varies minutely, with the changing composition. The crystallization peak of PLLA appears at 108°C, while that of PHB does not appear clearly. The DSC thermal properties are summarized in Table 9.2. The crystallization peak for the 9:1 blend appears at 86°C, and the crystallization temperature decreases as the PLLA/PHB ratio changes from 9:1 to 6:4. The decreasing crystallization temperature indicates a faster crystallization rate of PLLA in the blend than pure PLLA, based on the ability of PHB to recrystallize the PLLA matrix [38,41]. The melt temperature slightly decreases with the increasing PHB content. A small exothermal peak at 159°C appears below the melting point; it may originate from the imperfect crystals that form during the electrospinning process and is affected by the heating rate [45–47]. The melting peak of all blends and PLLA shows a small significant shoulder, which is ascribed to the presence of two crystal forms and recrystallization effects during the heating process [48]. The crystallinity of PLLA generally increases as PHB is incorporated using the solution method [27,38]. However, in this study, the crystallinity slightly decreases, which can

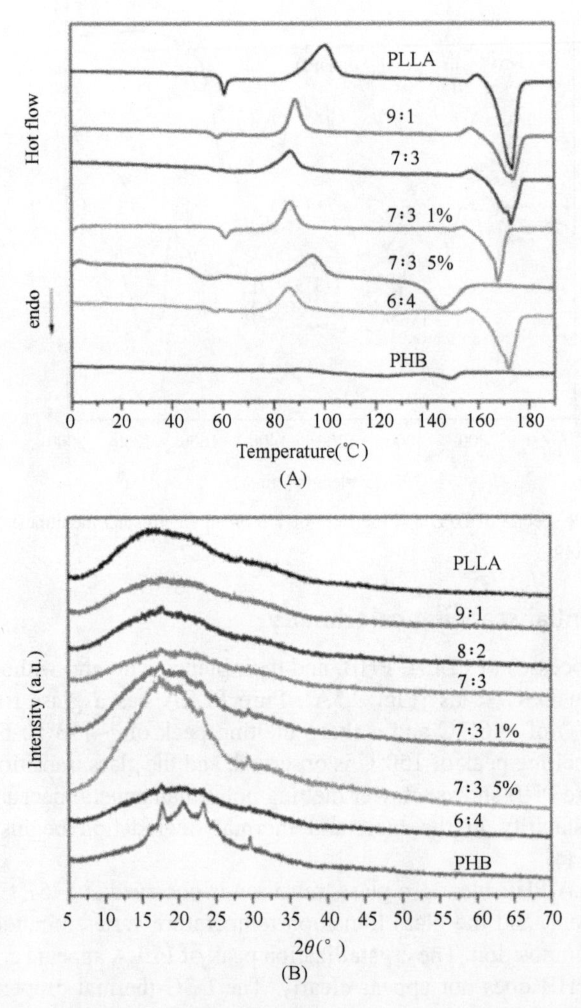

FIGURE 9.5 DSC scans (A) during the heating processes, and XRD (B) of PLLA, PHB, and PLLA/PHB blends with 0%, 1%, and 5% DPD [26].

be explained by two considerations: (1) melt electrospinning generally forms amorphous or semicrystalline structures [49]; and (2) molten polymers quenched at room temperature do not immediately crystallize [50]. Since the degree of tension has not reached the critical value of crystallization because of the rapid temperature drop, molecular chains in the melt are drawn by the electric force and solidify at the low temperature. Therefore, most of the molecular chains have a certain degree of orientation along the tensile direction, but the crystallinity is relatively low.

The T_g of PLLA/PHB with 1% DPD is slightly higher than that of PLLA/PHB. However, the T_g of PLLA/PHB with 5% DPD is significantly lower than

TABLE 9.2 DSC thermal results of electrospun mats [26].

Materials	T_g	T_c	T_m	$X_{c\ PLLA}$
PLLA	61.2	99.6	173.5	28.6%
9:1	58.1	88.3	174.0	21.2%
7:3	57.5	86.6	172.8	15.8%
6:4	57.2	86.1	171.8	15.7%
PHB	–	94.7	149.5	7.5%[a]
7:3/1%	60.8	86.4	168.1	10.6%
7:3/5%	54.0	95.0	145.2	3.1%

a X_c (%), calculated using ΔH_f of PHB of 146 ($J \cdot g^{-1}$) [27].

that of PLLA/PHB. Although no significant differences in the crystallization temperature are observed between the blend with 1% DPD and neat PLLA/PHB, the crystallization temperature of the blend with 5% DPD is increased. For the melting behavior, the addition of DPD significantly reduces the melting temperature of the blend by redistributing the crystals [51]. The melting peak becomes narrower, and the small shoulder that is observed for the neat PLLA/PHB blend is not observed for the blend with 1% DPD. This indicates that the formation of more stable crystals is promoted by adding DPD.

9.3.4 X-ray diffraction

The crystalline structure of the electrospun fibers is investigated by XRD, and the diffraction patterns are shown in Fig. 9.5B. An amorphous broadband is typically observed for the pure PLLA sample [52]. The weaker peak located at $2\theta = 31.0°$ is associated with the β-crystal, and the broad dispersion peaks at 16.6° and 19.0° are related to the (103) and (203) reflections, respectively, of the α-form crystal of PLLA [53,54]. The diffraction pattern for the electrospun film contains five peaks at 17.4°, 20.3°, 23.2°, 25.5°, and 29.6°, corresponding to the (110), (021), (111), (121), and (200) reflections of PHB, respectively [55]. The presence of crystalline PHB does not produce significant signals or increase the crystallinity of the PLLA matrix, in agreement with a slight decrease in the degree of crystallinity calculated from the DSC results. In the 6:4 blend, weak peaks are observed at $2\theta = 17.4°$, 23.2°, and 29.6°. These findings suggest that the crystal structure of PHB in the 6:4 electrospun membranes is altered; the crystal growth rate of PLLA is slower than that of PHB, which creates interactions between PLLA and PHB [55]. The 7:3 blend with 1% DPD shows a slight increase in the intensity of the peaks at $2\theta = 17.4°$, 20.3°, 23.2°, and 29.6° with respect to those for the pure 7:3 blend; this may be attributed to the high dispersion of DPD, which contributed to crystal formation.

9.3.5 Electron spin-resonance probe spectroscopy of polylactic acid (PLA)/polyhydroxybutyrate (PHB) electrospun mats

The last four sections have presented structural, morphological, and thermal information, which shows that special features exist for the PLLA/PHB blend fibers fabricated by melt electrospinning. The next two sections will describe the dynamic behavior of polymer molecules, namely their segmental mobility, as evaluated using a probe ESR method and DPD diffusion measurements, which are the most important processes for controlled drug release, especially for its short-term development. The molecular structural features, morphology, and physicochemical characteristics provide the necessary conditions, while the kinetics of diffusive transport in combination with the hydrolysis rate provides sufficient conditions for drug release in biodegradable fibers.

A series of ESR spectra for the (2,2,6,6-tetramethylpiperidin-1-yl)oxyl (TEMPO) radical probe encapsulated in the PLLA/PHB systems (microfibers and microparticles) are shown in Fig. 9.6. The shapes of all ESR curves indicate clearly that each inherent spectrum manifests the superposition of two individual spectra that belong to the two populations of radicals with two different correlation times, τ_1 and τ_2. Here, the intrinsic correlation time, τ_1, designates the status of the radical in the dense amorphous fields with slow rotation mobility, while τ_2 designates fast radical rotation in the less dense amorphous fields of the fibers. The latter fields are designated as the "soft" fields.

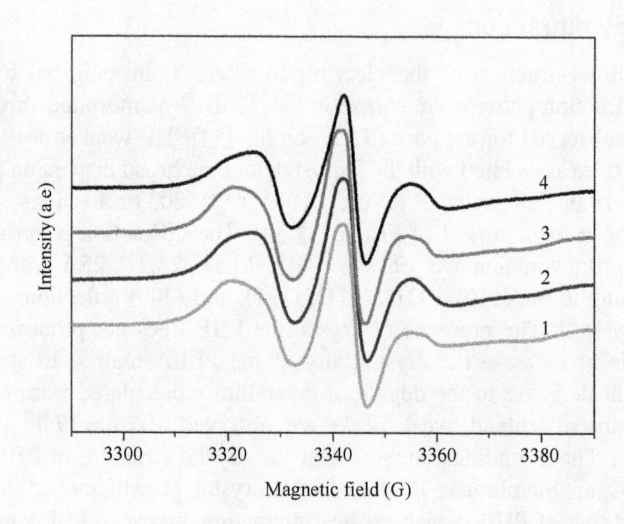

FIGURE 9.6 Probe ESR spectra of TEMPO radicals embedded in PLLA/PHB fibrillar systems: (1) 9: 1/5%; (2) 7: 3/5%; (3) 9: 1/1%; (4) 7: 3/1% [26].

The existence of two TEMPO populations in the amorphous phase of PHB and PLLA with distinctive rotation frequencies indicates the heterogeneous structure of the intercrystalline areas of the biopolymers. The state of the intercrystalline fraction could be approximated by a two-mode model, which is previously proposed for several semicrystalline polymers, such as poly-alkanoates, polylactides, and poly(ethylene terephthalate) [56,57].

Quantitative analysis of the ESR spectra was performed by evaluating the effective correlation times in accordance with Eq. (9.2). Additionally, the ratio of the intensities of the two first low-field peaks, I_+^1 and I_+^2, belong to slow and fast rotations, respectively (see Fig. 9.7); the activation energy of the

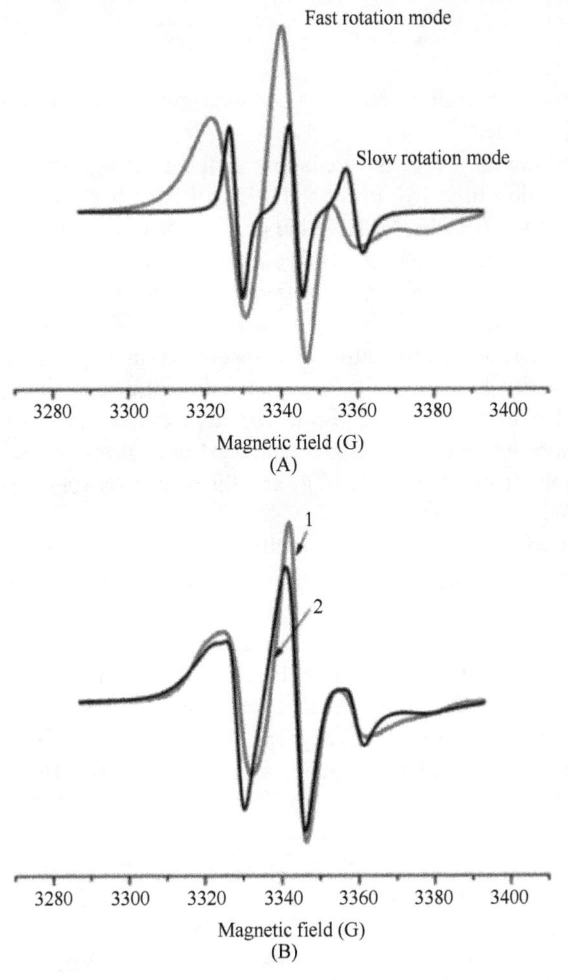

FIGURE 9.7 ESR spectrum deconvolution (#2 in Fig. 6.6): (A) the fast and slow components of ESR spectrum after deconvolution; (B) the comparison of the real initial spectrum (1) with the spectrum evaluated as a superposition (2) of two different modes shown in (A) [26].

TABLE 9.3 ESR spectral characteristics of radical TEMPO embedded in the PLLA/PHB mats[26].

Materials	Correlation time (10^9, s)	I_+^1/I_+^2	E_a (kJ·mol^{-1})
9:1/1%	6.2	0.55	42
9:1/5%	8.8	0.61	43 (20)
7:3/1%	3.6	0.48	36.5
7:3/5%	4.2	0.52	41.5 (13.5)

radical rotational mobility, E_a, is also measured. All characteristics are summarized in Table 9.3.

The differences in the peak intensities of the ESR spectra show that the effective correlation time, τ_C, in the 9:1 PLLA/PHB fibers exceeds the same characteristic of the 7:3 PLLA/PHB fibers, which also indicates the decline in segmental mobility in the amorphous fraction of PLLA, with an increase in the PLLA content. The concentration effect is quite explicable, especially considering that the T_g of PLLA, by more than 36°, exceeds the T_g of PHB. Hence, at room temperature, the former is in a glassy state where segmental mobility is essentially hindered, while the latter has the limited features of an elastic polymer; despite the high crystallinity, its segmental mobility is more intense. The relative standard deviations of the correlation times are 6%, the intensity ratio is 10%, and the activation energy of the probe mobility is 5%.

A comparison of the values for the I_+^1/I_+^2 ratio shows that the partition coefficients of the radicals in the dense and soft fields of the amorphous fraction of the polymer do not differ much. Nevertheless, the perfect fibrils show slightly higher intensity ratios (see Table 9.3) than the distorted PLLA/PHB systems (see Fig. 9.2H). This result points out the denser structural organization in the intercrystalline area in the well-formulated fibrillar. Here, it is appropriate to mention the work [58], where an analogous small discrepancy is shown between the meaning of the I_+^1/I_+^2 ratio for the ultrathin fibrils and the cold-rolled films of the pristine PHB.

In Table 9.3, the important characteristics of radical mobility are presented as energy activation values, calculated from the Arrhenius equation in the coordinates $\log_{10}(\tau_C)$ versus $1/T$ K. The Arrhenius plots for microparticles are not monotonic and show specific breakpoints that reflect a sharp change in the slope of the corresponding straight lines at ~57°C and ~39°C for the systems containing 5% DPD and with polymer ratios of 9:1 and 7:3,

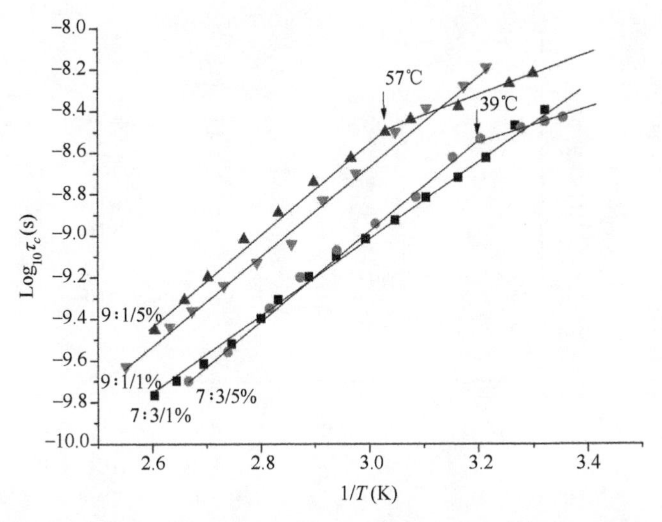

FIGURE 9.8 Semilogarithmic dependence of radical rotational mobility on reciprocal absolute temperature for PLLA/PHB systems. The arrows show the breakpoints expressed in Celsius degrees [26].

respectively. The decrease in the rotational activation energy is likely related to the acceleration of the rotation probe mobility near the glassy state transition in PLLA, as in the dominant component ($T_g = 58°C$ for the 9:1 ratio, see Table 9.2). For the 7:3 systems, PHB impacts the total mobility of the radical; thus, we could treat this mobility transition as the superposition of the segmental dynamics of both biopolymers. In any case, this temperature effect demands further investigation. Finally, all these transitions are observed for the mats formed with microparticles only at higher DPD contents (Fig. 9.8).

9.3.6 Impact of diffusion upon controlled drug release

Fig. 9.9A demonstrates the typical kinetic profiles of drug release from the PLLA/PHB fiber mats with DPD loading and different ratios of the polymer components. As for drug release from the PHB films [59], the kinetic profiles have two inherent sections characterized by different shapes for the mat: linear and nonlinear. In accordance with the proposed diffusion-kinetic model, the initial nonlinear section of each curve principally reflects the drug diffusion process, whereas the linear section corresponds to the kinetic process of partial loss of polymer weight because of the onset of the hydrolytic decomposition of PLLA and PHB ester groups. During hydrolysis, the encapsulated drug moves into aqueous surroundings, not only by a diffusion mechanism but also by surface degradation.

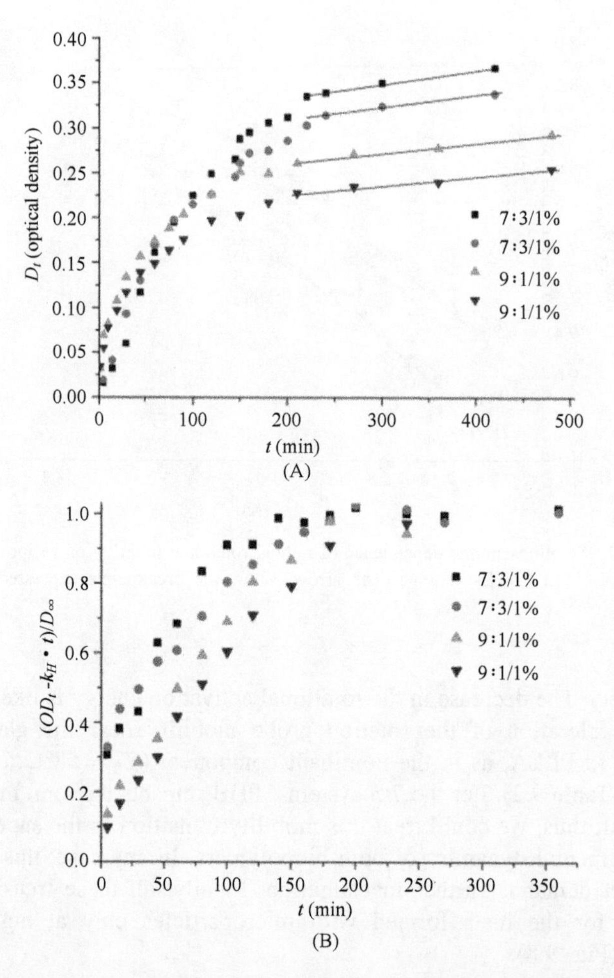

FIGURE 9.9 (A) Kinetic profiles of drug release from PLLA/PHB mats. (B) Diffusion impact upon the drug controlled release (DPD). Kinetic curves reflect drug diffusivity in the polymer phase [26].

For the macroscopic fibrillar membranes (the mats) formed by single filaments, by analogy with the monolithic nonporous film, the diffusion-kinetic equation can be written to reflect both diffusion and hydrolysis simultaneously as:

$$(\partial M_d/\partial t)/V = \partial C_d/\partial t = D_{eff}\left(\partial^2 G_d/\partial x^2\right) + k_h \tag{9.3}$$

where, C_d and G_d are the total concentration of the drug passing from the polymer into the external aqueous volume, V, and the concentration of the drug that can diffuse into the polymer, respectively; D_{eff} is the effective diffusion coefficient, which is independent of coordinates and time, and k_h is the constant of the zero-order hydrolytic reaction of PLLA and PHB ester groups on the filament surface.

The PLLA/PHB mats consist of disordered entangled fibers, and D_{eff} is determined for these mats by two consecutive processes; the drug diffusion mobility in the inherent fiber volume (D_f) and drug transport in the pores consisting of interfibrous space and filled by solvent (D_w). By describing the two-stage drug transport as a sequence of proper diffusion in the fiber and subsequent transfer in the interfibrous space, that is, modeling the diffusion mode of release as a two-layered medium in accordance with the Crank simplification [60], D_{eff} could be presented as:

$$L_M/D_{eff} = X_f/D_f + L_w/D_w \tag{9.4}$$

where, X_f and L_w are the average characteristic sizes of the drug diffusion path length in the proper fiber and the interfibrous space, respectively.

For the cylindrical fibers, X_f is the fiber diameter, while for L_w, in accordance with the Mackie -Mears equation [61], the correction for increasing the drug diffusion path based on its tortuosity is chosen as:

$$L_w = [(1+\varphi_f)/1-\varphi_f]L_M \tag{9.5}$$

where, φ_f is the volume fraction of polymer fibers. This correction was previously used to describe the drug diffusion in the PHB magnetic composites, with the magnetite nanoparticles embedded, forming the extended aggregates [62].

The diffusion equation for the cylindrical fibers loaded with the uniformly distributed drug is advanced by Crank [60]:

$$\begin{gathered}\partial G_d/\partial t = (1/r)D_f[\partial(r\partial G_d/\partial r)/\partial r] \\ \text{at} \quad 0 < r < X_f/2\end{gathered} \tag{9.6}$$

where, r is the coordinate of the radial diffusion; G_d, as in Eq. (9.3), denotes the concentration of the mobile fraction of the drug in the cylindrical fiber with the corresponding constant diffusion coefficient D_f, and X_f is the average diameter of the fiber. The initial and symmetrical boundary conditions corresponding to the drug desorption from the cylindrical fibers are the following: $G_f = G_f^0$ at $t = 0$ (at the initial time) and $G_f = 0$ at $r = R$ (at the fiber/solution interface).

In accordance with the research of J. Siepmann et al. [63], the solution of differential Eq. (9.6) makes it possible to obtain the dependence of the cumulative amount of the desorbed drug (M_t) on the time of release (t):

$$M_t/M_\infty = [16D_f/\pi]1/2t^{1/2} - [2D_f/X_f]t \tag{9.7}$$

where, M_∞ is the limiting value of M_t under the condition: $t \rightarrow \infty$.

For Eqs. (9.6) and (9.7), which both reflect drug transfer through the surface of the side walls and in a cylinder, the cylindrical fiber is at least five times longer than its radius [64]. This mathematical requirement is completely fulfilled for electrospun fibers of practically infinite length. Additionally, the latter equation is valid when $M_t/M_\infty \leqslant 0.4$. The combination of Eqs. (9.3) and (9.7) gives the final expression for drug release from cylindrical fibers that are simultaneously subjected to hydrolysis and diffusion:

$$M_t/M_\infty = [16D_f/\pi X_f]^{1/2}t^{1/2} + k_c t \quad (9.8)$$

where, $k_c = k_h - [4D_f/X_f^2]$. The positive sign in the equation shows that the condition, $k_h > [4D_f/X_f^2]$, is fulfilled for the PLLA/PHB fibers, which is confirmed in Fig. 9.9B. For a relatively short exposure time (~8 h) and given the S-mechanism of PLA and PHB hydrolysis under short-term controlled release [65], no changes are observed in the viscosity, molecular weight, or crystallinity. The experimental curves are consistent with that provided by Eq. (9.8), which provides a method to evaluate the drug diffusivity in the diffusion coordinates: $M_t/M_\infty \approx t^{1/2}$. Table 9.4 lists the drug release characteristics of the studied systems, including the diffusion coefficients (D_f) from Eq. (9.8). The low values of D_f satisfactorily agree with the same low values for diffusion mobility of DPD in the PHB microparticle [66].

The penultimate column in Table 9.4 shows the values of k_h, which were calculated from the curves presented in Fig. 9.9B using the last member of Eq. (9.8). The 9:1 PLLA/PHB system is more resistant to polymer hydrolysis than the 7:3 systems. The partial incompatibility of PLLA and PHB could increase the microscale phase separation surface of the single filaments, making them more accessible to attack by the hydrolytic agent.

The right column of Table 9.4 shows the mobile fraction of the drug encapsulated in the microfibrils, which are capable of diffusion. Simultaneous collation of the effective DPD diffusivities in combination with the values of the DPD mobile fractions shows that the rate of diffusion transport is approximately twice as high for the fibers with less PLLA (i.e., 7:3) than for

TABLE 9.4 Diffusion data for controlled drug release [26].

Materials	$D_f \cdot 10^{11}$ (cm$^2 \cdot$s^{-1})	$X_f \cdot 10^3$ (cm)	$k_h \cdot 10^6$ (s)	DPD ratio released via diffusion (%)
9:1/1%	4.9	3	1.9	6.4
9:1/1%	4.8	3	1.6	4.8
7:3/1%	9.9	2	8.3	16.2
7:3/1%	8.4	2	5.6	12.9

those in the 9:1 system. This effect is associated with the essential drop in fiber crystallinity, which is shown in Table 9.2. In accordance with the two-phase model for the amorphous areas of polymers, which is confirmed using ESR, the rest of the drug molecules are tightly encapsulated in the amorphous, dense areas and do not participate in diffusion transport. This immobilized fraction is only capable of drug release because of the surface destruction in accordance with a zero-order reaction. DPD release provides multiple bio-functions for biodegradable nanofibrous films, which show promise for use in biodegradable, small-diameter vascular grafts and other blood-contact implants and are candidates for drug-delivery systems for sustained delivery of pharmaceutical agents for applications related to cardiovascular disease.

9.4 Conclusion

In this study, the fiber membranes of PLLA/PHB blended with different concentrations of DPD are fabricated using solvent-free melt electrospinning to create controlled-release systems. The fibers without DPD had a smooth surface and a relatively uniform fiber diameter distribution. The fibers with DPD shows a rough surface and nonuniform diameters within a single fiber. The fibers with DPD include crazes that resembled lotus root crevices. The DSC results suggest that the crystallinity of PLLA slightly decrease with PHB incorporation; this is further confirmed using XRD. The addition of DPD significantly reduces the melting temperature of the blend. FTIR spectroscopy shows a transesterification reaction between PLLA and PHB and confirmed that the drug was immobilized on the fibers.

Furthermore, the dynamic behavior of the polymer molecules is evaluated by segmental mobility and drug diffusion. In the blend fibers, the segmental mobility of PLLA and PHB decline with increasing PLLA content. The values of I_+^1/I_+^2 indicate a denser structural organization of the intercrystalline area in the fibrils. The decrease in the rotation activation energy is likely related to the acceleration of the rotation probe mobility near the glassy-state transition. The consideration of drug release under short-term conditions shows that these nanofibers can be characterized by the diffusion mode of a sustained drug release profile. The 9:1 PLLA/PHB system is more resistant to polymer hydrolysis than the 7:3 system. The simultaneous collation of the effective DPD diffusivities, in combination with the DPD mobile fractions, shows that the rate of diffusion transport is approximately two times higher for the 7:3 PLLA/PHB fibers than for the 9:1 PLLA/PHB fibers.

References

[1] Uhrich KE, Cannizzaro SM, Langer RS, et al. Polymeric systems for controlled drug release. Chemical Reviews 1999;99(11):3181–98.

[2] Kong L, Ziegler GR. Fabrication of pure starch fibers by electrospinning. Food Hydrocolloids 2014;36:20–5.

[3] Langer R. Drug delivery and targeting. Nature 1998;392:5−10.
[4] Ali AQ, Kannan TP, Ahmad A, et al. In vitro genotoxicity tests for polyhydroxybutyrate − a synthetic biomaterial. Toxicology in Vitro 2008;22(1):57−67.
[5] You JW, Chiu HJ, Don TM. Spherulitic morphology and crystallization kinetics of melt-miscible blends of poly(3-hydroxybutyrate) with low molecular weight poly(ethylene oxide). Polymer 2003;44(15):4355−62.
[6] Santos D, Silva DM, Gomes PS, et al. Multifunctional PLLA-ceramic fiber membranes for bone regeneration applications. Journal of Colloid and Interface Science 2017;504:101−10.
[7] Oyama HT, Tanishima D, Ogawa R. Biologically safe Poly(l\r, l\r, -lactic acid) blends with tunable degradation rate: microstructure, degradation mechanism, and mechanical properties. Biomacromolecules 2017;18(4):1281−92.
[8] Badia JD, Reig-Rodrigo P, Teruel-Juanes R, et al. Effect of sisal and hydrothermal aging on the dielectric behavior of polylactide/sisal biocomposites. Composites Science and Technology 2017;149(8):1−10.
[9] Zeng J, Xu X, Chen X, et al. Biodegradable electrospun fibers for drug delivery. Journal of Controlled Release 2003;92(3):227−31.
[10] Xie G, Wang Y, Han X, et al. Pulsed electric fields on poly-L-(lactic acid) melt electrospun fibers. Industrial and Engineering Chemistry Research 2016;55(26):7116−23.
[11] Dalton PD, Klinkhammer K, Salber J, et al. Direct in vitro electrospinning with polymer melts. Biomacromolecules 2006;7(3):686−90.
[12] Rosendaal FR, Algra A. Dipyridamole and acetylsalicylic acid in the secondary prevention of stroke. Journal of the Neurological Sciences 1997;150(1):85−7.
[13] D'Ilario L, Francolini I, Martinelli A, et al. Dipyridamole-loaded poly(L-lactide) single crystals as drug delivery systems. Macromolecular Rapid Communications 2010;28(18−19):1900−4.
[14] Repanas A, Bader A, Klett A, et al. The effect of dipyridamole embedded in a drug delivery system made by electrospun nanofibers on aortic endothelial cells. Journal of Drug Delivery Science and Technology 2016;35:343−52.
[15] Punnakitikashem P, Truong D, Menon JU, et al. Electrospun biodegradable elastic polyurethane scaffolds with dipyridamole release for small diameter vascular grafts. Acta Biomaterialia 2014;10(11):4618−28.
[16] Besheli NH, Mottaghitalab F, Eslami M, et al. Sustainable release of vancomycin from silk fibroin nanoparticles for treating severe bone infection in rat tibia osteomyelitis model. ACS Applied Materials and Interfaces 2017;9(6):5128−38.
[17] Harnoy AJ, Buzhor M, Tirosh E, et al. Modular synthetic approach for adjusting the disassembly rates of enzyme-responsive polymeric micelles. Biomacromolecules 2017;18(4):18−28.
[18] Almeida A, Diening L. Water state effect on drug release from an antibiotic loaded polyurethane matrix containing albumin nanoparticles. International Journal of Pharmaceutics 2011;407(1):197−206.
[19] Wang Y, Pan J, Han X, et al. A phenomenological model for the degradation of biodegradable polymers. Biomaterials 2008;29(23):3393−401.
[20] Keshavarz P, Ayatollahi S, Fathikalajahi J. Mathematical modeling of gas−liquid membrane contactors using random distribution of fibers. Journal of Membrane Science 2008;325(1):98−108.
[21] Hopfenberg HB. In: Controlled release polymeric formulations, vol. 26. Washington, DC: American Chemical Society; 1976.

[22] Azwa ZN, Yousif BF, Manalo AC, et al. A review on the degradability of polymeric composites based on natural fibres. Materials and Design 2013;47(9):424–42.
[23] Li X, Liu Y, Peng H, et al. Effects of hot airflow on macromolecular orientation and crystallinity of melt electrospun poly(L-lactic acid) fibers. Materials Letters 2016;176:194–8.
[24] Liu Y, Li X, Ramakrishna S. Melt electrospinning in a parallel electric field. Journal of Polymer Science Part B: Polymer Physics 2014;52(14):946–52.
[25] Nagy ZK, Balogh A, Dravavoelgyi G, et al. Solvent-free melt electrospinning for preparation of fast dissolving drug delivery system and comparison with solvent-based electrospun and melt extruded systems. Journal of Pharmaceutical Sciences 2013;102(2):508–17.
[26] Cao K, Liu Y, Olkhov AA, Siracusa V, Iordanskii AL. PLLA-PHB fiber membranes obtained by solvent-free electrospinning for short-time drug delivery. Drug Deliv Transl Res. 2018;8(1):291–302.
[27] Arrieta MP, LópezMartínez J, López D, et al. Development of flexible materials based on plasticized electrospun PLA PHB blends: structural, thermal, mechanical and disintegration properties. European Polymer Journal 2015;73:433–46.
[28] Repanas A, Glasmacher B. Dipyridamole embedded in Polycaprolactone fibers prepared by coaxial electrospinning as a novel drug delivery system. Journal of Drug Delivery Science and Technology 2015;29:132–42.
[29] Budil DE, Lee S, Saxena S, et al. Nonlinear-least-squares analysis of slow-motion EPR spectra in one and two dimensions using a modified levenberg–marquardt algorithm. Journal of Magnetic Resonance, Series A 1996;120(2):155–89.
[30] Russian AI. Simulation of EPR spectra of the radical TEMPO in water-lipid systems in different microwave ranges. Biofizika 2011;56(3):407–17.
[31] Buchachenko AL, Vasserman AM. Stable radicals. Moscow: Khimiya; 1973.
[32] Asran AS, Seydewitz V, Michler GH. Micromechanical properties and ductile behavior of electrospun polystyrene nanofibers. Journal of Applied Polymer Science 2012;125(3):1663–73.
[33] Yoshioka T, Dersch R, Greiner A, et al. Highly oriented crystalline PE nanofibrils produced by electric-field-induced stretching of electrospun wet fibers. Macromolecular Materials and Engineering 2010;295(12):1082–9.
[34] Liu Y, Liu Y, Lee J, et al. Ultrafine formation of optically transparent polyacrylonitrile/polyacrylic acid nanofibre fibrils via electrospinning at high relative humidity. Composites Science and Technology 2015;117:404–9.
[35] Ling XY, Spruiell JE. Analysis of the complex thermal behavior of poly(L-lactic acid) film. II. Samples crystallized from the melt. Journal of Polymer Science Part B: Polymer Physics 2010;44(23):3378–91.
[36] Zeng J, Yang L, Liang Q, et al. Influence of the drug compatibility with polymer solution on the release kinetics of electrospun fiber formulation. Journal of Controlled Release 2005;105(1–2):43–51.
[37] Hu Y, Sato H, Zhang J, et al. Crystallization behavior of poly(l-lactic acid) affected by the addition of a small amount of poly(3-hydroxybutyrate). Polymer 2008;49(19):4204–10.
[38] Furukawa T, Sato H, Murakami R, et al. Structure, dispersibility, and crystallinity of poly(hydroxybutyrate)/poly(l-lactic acid) blends studied by FT-IR microspectroscopy and differential scanning calorimetry. Macromolecules 2005;38(15):6445–54.
[39] Sato H, Murakami R, Padermshoke A, et al. Infrared spectroscopy studies of CH···O hydrogen bondings and thermal behavior of biodegradable poly(hydroxyalkanoate). Macromolecules 2004;37(19):7203–13.

[40] Padermshoke A, Katsumoto Y, Sato H, et al. Melting behavior of poly(3-hydroxybutyrate) investigated by two-dimensional infrared correlation spectroscopy. Spectrochimica Acta Part A Molecular and Biomolecular Spectroscopy 2005;61(4):541–50.

[41] Zhang M, Thomas NL. Blending polylactic acid with polyhydroxybutyrate: the effect on thermal, mechanical, and biodegradation properties. Advances in Polymer Technology 2011;30(2):67–79.

[42] Borges CPF, Tabak M. Spectroscopic studies of dipyridamole derivatives in homogeneous solutions: effects of solution composition on the electronic absorption and emission. Spectrochimica Acta Part A Molecular Spectroscopy 1994;50(6):1047–56.

[43] Oliveira MS, Agustinho SC, Plepis AMD, et al. On the thermal decomposition of dipyridamole: thermogravimetric, differential scanning calorimetric and spectroscopic studies. Spectroscopy Letters 2006;39(2):145–61.

[44] Kunioka M, Saito T, Doi Y, et al. Biodegradation of microbial copolyesters: poly(3-hydroxybutyrate-co-3-hydroxyvalerate) and poly(3-hydroxybutyrate-co-4-hydroxybutyrate). Macromolecules 1990;23(1):26–31.

[45] Ling X, Spruiell JE. Analysis of the complex thermal behavior of poly(L-lactic acid) film. I. Samples crystallized from the glassy state. Journal of Polymer Science Part B: Polymer Physics 2010;44(22):3200–14.

[46] Zhang J, Duan Y, Sato H, et al. Crystal modifications and thermal behavior of poly(l-lactic acid) revealed by infrared spectroscopy. Macromolecules 2005;38(19):8012–21.

[47] Zong XH. Structure and process relationship of electrospun bioabsorbable nanofiber membranes. Polymer 2002;43(16):4403–12.

[48] Monticelli O, Bocchini S, Gardella L, et al. Impact of synthetic talc on PLLA electrospun fibers. European Polymer Journal 2013;49(9):2572–83.

[49] Brown TD, Dalton PD, Hutmacher DW. Melt electrospinning today: an opportune time for an emerging polymer process. Progress in Polymer Science 2016;56:116–66.

[50] Furuhashi Y, Imamura Y, Jikihara Y, et al. Higher order structures and mechanical properties of bacterial homo poly(3-hydroxybutyrate) fibers prepared by cold-drawing and annealing processes. Polymer 2004;45(16):5703–12.

[51] Karpova SG, Ol'Khov AA, Shilkina NG, et al. Influence of drug on the structure and segmental mobility of poly(3-hydroxybutyrate) ultrafine fibers. Polymer Science Series A 2017;59(1):58–66.

[52] Channuan W, Siripitayananon J, Molloy R, et al. The structure of crystallisable copolymers of l-lactide, ε-caprolactone and glycolide. Polymer 2005;46(17):6411–28.

[53] El-Hadi AM, Mohan SD, Davis FJ, et al. Enhancing the crystallization and orientation of electrospinning poly (lactic acid) (PLLA) by combining with additives. Journal of Polymer Research 2014;21(12):1–12.

[54] Nisha SK, Asha SK. Random copolyesters containing perylene bisimide: flexible films and fluorescent fibers. ACS Applied Materials and Interfaces 2014;6(15):12457–66.

[55] Abdelwahab MA, Flynn A, Chiou BS, et al. Thermal, mechanical and morphological characterization of plasticized PLA–PHB blends. Polymer Degradation and Stability 2012;97(9):1822–8.

[56] Kamaev PP, Aliev II, Iordanskii AL, et al. Molecular dynamics of the spin probes in dry and wet poly(3-hydroxybutyrate) films with different morphology. Polymer 2001;42(2): 515–20.

[57] Di Lorenzo ML, Gazzano M, Righetti MC. The role of the rigid amorphous fraction on cold crystallization of poly(3-hydroxybutyrate). Macromolecules 2012;45(14):5684–91.

[58] Staroverova O, Karpova S, Iordanskii A, et al. Comparative dynamic characteristics of electrospun ultrathin fibers and films based on poly(3-hydroxybutyrate). Chemical Technology 2016;10:151–8.
[59] Iordanskii AL, Rogovina SZ, Kosenko RY, et al. Development of a biodegradable polyhydroxybutyrate-chitosan-rifampicin composition for controlled transport of biologically active compounds. Doklady Physical Chemistry 2010;431(2):60–2.
[60] Crank J. The mathematics of diffusion. 2nd ed. Oxford: Clarendon Press; 1975.
[61] Mackie JS, Meares P. The diffusion of electrolytes in a cation-exchange resin membrane. I. Theoretical. Proceedings of the Royal Society A: Mathematical, Physical and Engineering Sciences 1955;232(1191):498–509.
[62] Bychkova AV, Iordanskii AL, Kovarski AL, et al. Magnetic and transport properties of magneto-anisotropic nanocomposites for controlled drug delivery. Nanotechnologies in Russia 2015;10(3–4):325–35.
[63] Siepmann J, Siepmann F. Mathematical modeling of drug delivery. International Journal of Pharmaceutics 2008;364(2):328–43.
[64] Iordanskii AL, Ol'khov AA, Karpova SG. Influence of the structure and morphology of ultrathin poly(3-hydroxybutyrate) fibers on the diffusion kinetics and transport of drugs. Polymer Science Series A 2017;59(3):352–62.
[65] Bonartsev AP, Boskhomodgiev AP, Iordanskii AL, et al. Hydrolytic degradation of poly(3-hydroxybutyrate), polylactide and their derivatives: kinetics, crystallinity, and surface morphology. Molecular Crystals and Liquid Crystals 2012;556(1):288–300.
[66] Bonartsev AP, Livshits VA, Makhina TA, et al. Controlled release profiles of dipyridamole from biodegradable microspheres on the base of poly(3-hydroxybutyrate). Express Polymer Letters 2007;1(12):797–803.

Index

Note: 'Page numbers followed by "f" indicate figures and "t" indicate tables'.

E

R

S

T

U

V

X

Y